V.I. Arnol'd

D1384827

HUYGENS AND BARROW, NEWTON AND HOOKE

Pioneers in mathematical analysis and catastrophe theory from evolvents to quasicrystals

Translated from the Russian by Eric J.F. Primrose

1990   Birkhäuser Verlag
Basel · Boston · Berlin

Author's address:
V.I. Arnol'd
Steklov Mathematical Institute
Vavilova 42
117 966 Moscow GSP-1

Originally published as: Gyuigens i Barrou, N'yuton i Guk.
Nauka FML, Moscow 1989

The colour illustrations were prepared by J.H. Lowenstein, Professor of Physics at New York University, to accompany Professor Arnol'd's James Arthur Lecture, "Quasicrystallic Symmetry Generated by the Time Evolution in Dynamical Systems", November 2, 1988.

Library of Congress Cataloging-in-Publication Data
Arnol'd, V.I. (Vladimir Igorevich), 1937–
[Giuigens i Barrou, N'iuton i Guk. English] Huygens and Barrow, Newton and Hooke : pioneers in mathematical analysis and catastrophe theory from evolvents to quasicrystals / V.I. Arnol'd ; translated from the Russian by Eric J.F. Primrose.
Translation of: Giuigens i Barrou, N'iuton i Guk.
ISBN 0-8176-2383-3 (U.S. : alk. paper)
1. Mathematical analysis – History – 17th century. 2. Mathematical physics – History – 17th century. I. Title.
QA300.A73513   1990
515'.09'032 – dc20   90-38298

Deutsche Bibliothek Cataloging-in-Publication Data
Arnol'd, Vladimir I.:
Huygens and Barrow, Newton and Hooke : pioneers in mathematical analysis and catastrophe theory from evolvents to quasicrystals / V.I. Arnol'd. Transl. from the Russ. by Eric J. F. Primrose. – Basel ; Boston ; Berlin : Birkhäuser, 1990
   ISBN 3-7643-2383-3

© 1990 for the English edition
Birkhäuser Verlag Basel
Printed in Germany on acid-free paper
ISBN 3-7643-2383-3
ISBN 0-8176-2383-3

CONTENTS

HUYGENS AND BARROW, NEWTON AND HOOKE

The year 1987 was the 300th anniversary of the publication of Newton's *Principia*, the book that laid the foundations of the whole of modern theoretical physics. Properly speaking, theoretical physics began with this book. At almost the same time mathematical analysis began. The first publication on analysis was in 1684, and it was due not to Newton, since he did not publish his discoveries in this field, but to Leibniz.

When we speak* of the contents of the *Principia* we should consider how the book was written, from what it arose, what problems were solved, when analysis was created, for what it was created, why it was so called, and where its basic ideas came from, for example, why in analysis we speak of functions and so on.

All these questions refer to the time of Newton at the end of the 17th century, when a whole galaxy of brilliant mathematicians

---

* The present book is an extended version of a lecture given on 25 February 1986 at the opening of the students' lecture series of the Moscow Mathematical Society. The author is grateful to A. Yu. Vaintrob, who passed on his record of this lecture, and also to V. L. Ginzburg, A. P. Yushkevich and G. K. Mikhailov for useful remarks. The lecture is supplemented by material from the articles "Three hundred years of mathematical science" (Priroda, 1987, No. 8, pp. 5–15) and "Kepler's second law and the topology of Abelian integrals" (Kvant, 1987, No. 12, pp. 17–21).

The author acknowledges the hospitality of New York University, where he read the 1988 series of the James Arthur Lectures on Time and its Mysteries, partially incorporated in this book, and the help of John H. Lowenstein, who produced the computer-made pictures for this lecture, partially reproduced in this book.

was working. The subsequent development of mathematics has completely overshadowed their achievements, hence the grandiose discoveries of those times seem to us from a distance to be less than they actually were. Among these mathematicians, apart from the best known, Descartes, Pascal and Fermat, who preceded Newton and Leibniz, and Johann Bernoulli, who was working a little later, we must mention Barrow, the direct predecessor and teacher of Newton, and Huygens, who solved the same problems as Newton and Leibniz, but usually somewhat outstripping them, even without any analysis.

The mathematical discoveries of Huygens had a strange fate. Most of them entered analysis not in his lifetime, but much later, and mainly thanks to the work of other mathematicians (for example Hamilton, who worked more than 100 years later). These results have now entered science in the form of symplectic geometry, the calculus of variations, optimal control, singularity theory, catastrophe theory,... We are only just learning about some of them. For example, it has recently become clear (a lecture of Bennequin *(1)** to the Bourbaki seminar) that the first textbook on analysis, written by l'Hôpital from the lectures of Johann Bernoulli, contains a representation of the manifold of irregular orbits of the Coxter group H_3 (generated by reflexions in the planes of symmetry of an icosahedron). This representation appears there not in connection with the group of symmetries of the icosahedron, but as a result of investigations of evolvents** of plane curves with a point of inflexion, investigations very close to those of Huygens (and possibly even carried out by him, although the first publication was apparently due to

---

*   References of this kind refer to the Notes at the end of the book.

**   (Translator's note.) In agreement with the author, the term "evolvent" has been used in this edition instead of the equivalent "involute", which is also commonly found in the literature.

l'Hôpital). Illustrations appearing in recent works on the connection between the icosahedron and singularities of evolutes and evolvents and, it should be said, obtained by modern mathematicians not without difficulty and even with the help of computers, were already known at that time.

We shall return to evolvents, and I am now speaking about the history of Newton's *Principia* and about the contents of the main part of this book. Essentially this book was written to solve a one-off problem. Although it contains, of course, the so-called three laws of Newton and a great quantity of other material, nevertheless it was actually written in less than a year only in order to present the solution of this one problem, namely the problem of motion in a force field inversely proportional to the square of the distance from an attracting centre.

The first part of the story is the history of where this problem came from, why Newton began to take it up, and what he proved, properly speaking, in this direction. This is the story of Newton and Hooke.

CHAPTER 1.
THE LAW OF UNIVERSAL GRAVITATION

§1. Newton and Hooke

The name of Newton and his enormous contribution to both mathematics and physics are well known. He was born in 1642, in the year of Galileo's death, and died in 1727. The work of Newton in the field of the theory of gravitation became famous in continental Europe thanks to Voltaire, who paid a visit to England in the last years of Newton's life and propagated the law of universal gravitation, which made a great impression on him. Voltaire informed the world about the famous apple, which Newton's niece Catherine Barton had told him about *(2)*.

Robert Hooke was an older contemporary of Newton, though much less well known. He was born in 1635 and died in 1703. Hooke was a poor man and began work as an assistant to Boyle (who is now well known thanks to the Boyle – Mariotte law discovered by Hooke *(3)*). Subsequently Hooke began working in the recently established Royal Society (that is, the English Academy of Sciences) as Curator. The duties of Curator of the Royal Society were very onerous. According to his contract, at every session of the Society (and they occurred every week except for the summer vacation) he had to demonstrate three or four experiments proving the new laws of nature.

Hooke held the post of Curator for forty years, and all that time he carried out his duties thoroughly. Of course, there was no condition in the contract that all the laws to be demonstrated had to be devised by him. He was allowed to read books, correspond with other scientists, and to be interested in their discoveries. He was only required to verify whether their statements were true and to convince the members of the Royal Society that some law was reliably established. For this it was necessary to prove this law experimentally and demonstrate the appropriate experiment. This was Hooke's official activity.

In the line of duty Hooke was interested in all discoveries in natural science by others, but it also fell to him to make discoveries. Towards the end of his life he counted 500 laws that he had discovered. It needs to be said that the numerous discoveries of Hooke form the basis of modern science. Very many of them were discovered more or less in parallel with other scientists, hence very often now laws discovered by Hooke are known, but attributed to other people. As a result the law of elasticity (the force is proportional to the extension) bears the name of Hooke, but his other discoveries bear other people's names. For example, Hooke discovered the cellular structure of plants. He improved the microscope and was the first to observe that plants consist of cells. He scrutinized various objects in the microscope and everything he saw he sketched. It is clear that as he saw new things in the microscope he quickly made new discoveries. Hooke personally engraved pictures which he saw in the microscope, and even published on the basis of this a book "Micrography", which later led Leeuwenhoek to his famous biological discoveries.

At that time it was easy to carry out fundamental discoveries, and large numbers of them were carried out. Huygens, for example, improved the telescope, looked at Saturn and discovered its ring, and Hooke discovered the red spot on Jupiter. At that time discoveries were not unusual events, they were not registered, not patented, as they are now, they were quite an everyday occurrence. (This was the case not only in the natural sciences. Mathematical discoveries at that time also poured forth as if from a horn of plenty *(4)*.)

12

But Hooke never had enough time to dwell on any of his discoveries and develop it in detail, since in the following week he needed to demonstrate new laws. So in the whole manifold of Hooke's achievements his discoveries appeared somewhat incomplete, and sometimes when he was in a hurry he made assertions that he could not justify accurately and with mathematical rigour.

One of the discoveries to which Hooke pretended was that of the wave nature of light. (That light consists of waves was asserted by Huygens at about the same time as Hooke.) Hooke based his conclusions on the study of colours of thin films (soap bubbles, for example). He assumed that the interference of light in soap films proves its wave structure. This produced the first conflict between Hooke and Newton.

Newton also took up the problem of light. He broke down white light into the components of the rainbow, determined the colours of the Sun's spectrum and thereby laid the foundations of modern spectroscopy, a science that is rather wave-theoretical. Nevertheless, Newton held to another theory and assumed that light consists of moving particles. Sound consists of waves, so sound can bend round obstacles (it can be heard even if the source is hidden by a hill, so essentially a hill is not an obstacle for sound), but light cannot bend round obstacles, we cannot see behind a hill, so light cannot consist of waves.

Despite the assertion that light consists of particles, Newton was the first to measure the wavelength of light. He did this in the following way. If we put a lens on glass and shine a light from above (Fig. 1), then the lengths of the paths of light rays that meet in one point will be different and, depending on whether the difference of the lengths is or is not an integer number of wavelengths, the rays will reinforce each other or cancel. Therefore, looking at the glass from above, we can see rings consisting of points of equal illumination (these are called Newton's rings, although Hooke discovered them). It is important that the thickness of the wedge of air between the lens and the glass is proportional to the *square* of the distance from the point of contact. Thanks to this the radii of the rings turn out to be proportional to the square root of the product of the wavelength and the radius of curvature of the lens. Because of this the radii of the rings are not as small as the wavelength, and the rings can be observed. By measuring these rings we can find the wavelength of light, which Newton did. But how did he calculate the wavelength if he did not believe in the wave nature of light? We are concerned with the specific character of Newton's theory of light. He assumed that particles of light fly in space not uniformly, but at the time

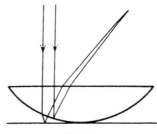

Fig. 1.
Formation of Newton's rings

of motion they experience periodic attacks (fits – something like modern ideas about internal degrees of freedom of particles). Thus he measured the distance between the positions of particles under two neighbouring attacks.

Thus, disagreements arose between Newton and Hooke. They could possibly have been avoided if there had not been aggravating circumstances. Newton lived in Cambridge and Hooke in London, and they were in correspondence mainly through Oldenburg, the secretary of the Royal Society. To all appearances, Oldenburg's character was not very good, and he took great pleasure in stirring up trouble between people. As a result, and because of their difference of opinion on the nature of light, relations between Newton and Hooke deteriorated completely. But after some time Oldenburg died (we shall return to him when we talk about analysis), and Hooke wrote a conciliatory letter to Newton. This letter written by Hooke on 24 November 1679 was essentially the start of the history of the law of universal gravitation *(5)*.

The purport of Hooke's conciliatory letter to Newton was the suggestion of joint work. Hooke recognized the famous achievements of Newton and suggested that they discuss and verify experimentally all possible ideas and theories. Hooke suggested to Newton, in particular, that Newton express his thoughts about several of Hooke's conjectures and promised not to be offended by criticism and, giving up their old dissensions, work together on the study of nature. In this letter Hooke informed Newton of the latest physical and mathematical news. One item of news was the latest theory of planetary motions, which had arrived from the continent of Europe. According to this theory it was assumed that whirlwinds are constantly raging in outer space, which carry the planets along, support them, and because of this compel them to rotate about the Sun. Another theory was Hooke's conjecture about attraction. In this letter he did not refer to it in detail, but only asked Newton what he thought about this conjecture. One more conjecture of Hooke was the law of oscillation of elastic bodies. In this letter Hooke discussed the new measurements of the length of a meridian (and consequently the radius of the Earth) by the French explorer Picard.

Newton answered very quickly – in four days. This remarkable letter writen by Newton on 28 November 1679 begins with an admission that he was finishing with philosophy and had recently been concerned with other matters. Apparently age was telling (Newton was already 37, and this was the age when it becomes difficult to be interested in mathematics and in other branches of philosophy). "I did not... so much as hear...", writes Newton, "of your hypotheses of compounding the celestial motions of the planets... though these no doubt are well known to the philosophical world... My affection to philosophy being worn out, so that I am almost as little concerned about it as one tradesman uses to be about another man's trade or a countryman about learning...".

The word "philosophy" at that time meant all the exact sciences. Physics was then called natural philosophy. The other

15

matters about which Newton wrote consisted, to all appearances, in his passion for alchemy. (Apparently he did not count this as philosophy, although the aim of this science consisted in finding the philosopher's stone.) Newton had a large chemical (or, if you like, alchemical) laboratory, and having worked intensively between the ages of 20 and 30 years in mathematics and physics and having done a great deal there, he was mainly concerned in obtaining gold. He collected a large number of alchemical recipes, preserved from the Middle Ages, and intended to manufacture gold in accordance with the instructions contained in them. The efforts he expended on this were significantly greater than those that went into the creation of his mathematical and physical works, but he did not derive any useful result. It is true that Newton was not convinced of this at times. It is said that in his note-books (and he wrote out his experiments in detail, describing what he mixed with what, and what results he obtained, so that if he obtained gold by chance, he could reproduce this process) there is an entry in which, after a detailed description of the actions he had carried out, he discussed the result: "Terrible stink. It seems, I am close to the target".

§2. The problem of falling bodies

Let us return to Newton's letter. He writes further that, although he had decided at such a venerable age not to compete with young minds, he could present a problem that seemed to him worthy of such a fine experimenter as Hooke. This was the problem of verifying the theory of Copernicus. As Copernicus asserted, the Earth moves around the Sun, and also rotates about its own axis. Newton suggested verifying the second assertion experimentally. In fact, according to Galileo's invariance principle, it is impossible to discover uniformly rectilinear motion in itself, but in principle a rotation could still be observed. There-

16

fore, said Newton, in order to convince those who do not believe in the theory of Copernicus (recognized by the Catholic Church, for example, only in 1837) it was necessary to test it experimentally. Apparently Newton was the first to pose the problem of an experimental proof of the rotation of the Earth. Moreover, in posing this problem to Hooke, Newton suggested a method that would, in principle, make it possible to do this.

Newton's suggestion was as follows. If the Earth rotates, then objects falling freely from a great height will be deflected from the vertical. Therefore it is sufficient to measure the deflection of the fall of heavy balls from the vertical direction (established by means of a plumb-line), in order to discover the rotation of the Earth.

In fact, Newton says in this letter, imagine that we are looking at the Earth from the North Pole and are seeing the equator and a mountain, or better a tower, from which freely falling balls have been dropped, initially at rest with respect to the tower (Fig. 2). Suppose that Copernicus is right, and the Earth rotates from west to east. An ignoramus would think, writes Newton, that then, while the ball is falling, the Earth under it would turn to the east and the ball would fall to the west of the place over which it was originally.

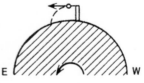

Fig. 2.
A trajectory in a non-rotating space

But this opinion, which is often advanced as an objection to the theory of Copernicus, is quite wrong. The mistake is that at the instant when it is dropped the ball has a non-zero initial speed with respect to the "fixed" system of reference, directed to the east. Moreover, the ball is above the Earth, so its speed is greater than the speed of points on the surface of the Earth. But the speed of the ball in the horizontal direction does not change during its fall, so it travels a greater distance in the easterly

17

direction than the point of the surface over which it was. Thus, the ball should fall not to the west, but to the east of this point.

If the balls are dropped not on the equator, but at our latitude, then the effect will be somewhat smaller, but nevertheless, says Newton, it should be possible to discover it. Of course, this effect is very small, so Newton advises doing the following. Under the point from which it is dropped and strictly from the plumb-line it is necessary to put a "steel" in the direction from north to south and to drop possibly heavier balls, having suspended them on a thread and burning it through in order to avoid unwanted initial jolts. Then, if we drop a ball sufficiently many times and calculate how many times the ball, on striking the steel, flies off to the east, and how many times to the west, we can, by comparing these two numbers, determine whether one can observe a subtle effect of deflecting to the east or not.

In his remarkable letter to Hooke, Newton touched on one more question. He discussed how a ball would move after reaching the surface if there were a shaft in the Earth (that is, the ball moves through the Earth without meeting any resistance). Newton assumed that the ball would then describe a spiral, and for clarity drew this spiral in the letter (Fig. 3).

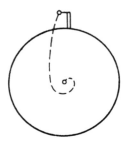

Fig. 3.
A trajectory inside the Earth according to Newton

Hooke read Newton's letter to the session of the Royal Society on 4 December 1679. This caused a lively discussion, in which many scientists took part. Everyone debated animatedly whether it was actually possible to observe the phenomenon described by Newton and on what side the balls must be deflected. For example, The Royal Astronomer Flamsteed came forward, as laid down in the protocol of the Society, with the statement that this effect had been known in the artillery for a long time. Namely, in Flamsteed's opinion, a shot falls back into the muzzle at an angle of elevation of 87° (apparently for this reason even now the stops of anti-aircraft guns do not allow the barrel to be raised to an angle greater than 87°). This, in Flamsteed's opinion, testifies to the rotation of the Earth, because otherwise the dangerous angle would be 90°. In other words, Flamsteed suggested a slight modification of Newton's suggestion. Instead of dropping the balls downwards, Flamsteed suggested shooting cannon balls vertically upwards and seeing whether they would fall back.

Hooke came forward at the next session on 11 December, making some critical remarks about Newton's arguments, to which Newton, who could not bear the slightest criticism, replied on 13 December with a long letter containing a lengthy discussion of the question and clearly showing that at the time Newton did not know what the trajectory of the ball should look like*.

Firstly, Newton's theory is very incomplete. It is necessary to take into account the fact that the direction of the vertical – the direction to the centre of the Earth – changes during the motion of the ball, so the force of gravity at different points of the trajectory is directed differently. This leads to the fact that a ball moving to the east will experience an influence that deflects it back to the west. So although the ball nevertheless falls to the east of a point of the plumb-line, the resulting deflection will be less than Newton predicted.

If, being provided with our modern knowledge, we carry out all the calculations accurately, we see that the true effect is 2/3 of the deflection that should be obtained according to Newton (6). Thus, the shift to the east on account of the difference in the

---

*   This letter contains among other mistakes an impossible picture of an orbit in a field whose potential is proportional to the distance from the centre: the angle between the pericentre and the apocentre is 120° (it should belong to the interval between $\pi/2$ and $\pi/\sqrt{3}$) and the orbit is clearly asymmetric.

distances to the centre of the Earth and the shift to the west caused by the difference in the direction of the force of gravity are quantities of the same order, so Newton's qualitative argument is altogether false. If these two effects – deflection to the east and deflection to the west – had a slightly different relation, the qualitative picture would be different.

Secondly, Hooke rightly observed that in the northern hemisphere a ball is deflected not only to the east, but also to the south. Moreover, he asserted (for incomprehensible reasons) that at our latitudes the deflection to the south is even greater than that to the east.

Finally, Hooke made a remark referring to the trajectory of motion of a ball inside the Earth. He says that the spiral drawn by Newton causes him some doubt. In his opinion, inside the Earth approximately the same will happen as under an oscillation of a pendulum on a string, and if a ball moves freely inside the Earth without experiencing resistance, then its trajectory will be closed

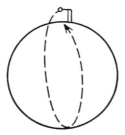

Fig. 4.
Trajectories inside the Earth according to Hooke

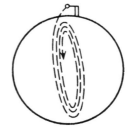

Fig. 5.
Taking account of air resistance according to Hooke

and remind us of an ellipse (Fig. 4), and a spiral can be obtained only by taking account of air resistance. But in this case the spiral obtained is not the same as Newton's, not making one circuit, but slowly winding, with a large number of rotations (Fig. 5).

In fact, if we solve this problem by means of our modern methods, we see that inside the Earth there acts not the law of universal gravitation but Hooke's law – the force of attraction is directly proportional to the distance from the centre of the Earth. Therefore inside the Earth the trajectory of a ball is the same as under elastic oscillations (or as with a pendulum), that is, it is elliptic.

In criticizing Newton, Hooke was not confined to theoretical arguments, and decided to carry out an experimental verification. He reported on his results to the Royal Society of 18 December. He organized the experiments somewhat differently and dropped the balls not on a "steel" situated below a layer of water, as Newton suggested, but in "a box filled by the tobacco pipe clay", which should weaken the shock force. On the box there was drawn a network of thin lines with centre under the point of suspension so as to determine from the trace of the ball not only the deflection to the west or east, but also in the north-south direction. The balls were dropped first in the open air and later in a cathedral from a height of about 9 metres with the doors and windows carefully closed, in order to protect the ball from the harmful effects of draughts. If everything is properly calculated, taking account of turbulence, then it is clear that with such a small height no effect could be observed (the theoretical deflection is 0.3 mm).

But Hooke was a very skilful experimenter. Till then this experiment had not been achieved, but Hooke "achieved" it. Hooke informed the Royal Society that in each of three trials a ball was deflected to the south-east by at least a quarter of an inch. Apparently he did not have any command of statistical analysis, and the number of trials was not large enough. In addition, most probably he did not verify the resulting deflection at the corresponding significance level and regarded the phenomenon as established, even though nothing had been clearly proved. At the beginning of 1680 Hooke repeated his experiments, again "successfully". He informed Newton of his results in a letter sent on 6 January.

§3. The inverse square law

Apart from the account of his experiments, this letter of Hooke contained the following important words: "My supposition is that the attraction always is in duplicate proportion to the distance from the centre reciprocal, and consequently that the velocity will be in a subduplicate proportion to the attraction, and consequently as Kepler supposes reciprocal to the distance... Mr. Halley, when he returned from St. Helena, told me that his pendulum at the top of the hill went slower than at the bottom, which he was much surprised at, and could not imagine a reason. But I presently told him that he had solved me a query I had long desired to be answered... and that was to know whether the gravity did actually decrease at a greater height from the centre... what I mentioned concerning the descent within the body of the Earth... not that I really believe there is such [inverse proportional to the squared distance] an attraction to the very centre of the Earth, but on the contrary I rather conceive that the more the body approaches the centre, the less will it be urged by the attraction, possibly somewhat like the gravitation on a pendulum or a body moved in a concave sphere where the power continually decreases the nearer the body inclines to a horizontal motion... But in the celestial motions the Sun, Earth or central body are the cause of the attraction, and though they cannot be supposed mathematical points yet they may be conceived as physical, and the attraction at a considerable distance may be computed accordingly to the former proportion [inverse square] as from the very centre...".

This inverse square law is apparently the Hooke theory of attraction that he mentioned in the first letter and, in Hooke's opinion, it was necessary to take it into account when studying the fall of a body both outside the Earth and inside. It is true, wrote Hooke, that inside the law is different, of course, since the layers traversed by the body will attract it in different directions. Therefore the law of motion inside is apparently similar to that observed in elastic oscillations. Hooke also wrote that when he was studying these laws of force he tried to determine the forms of the orbits in which bodies must move. He obtained the result that inside the Earth the orbits will be roughly the same as for oscillations of a pendulum, but outside, when there is only one attracting centre, the body will move along a curve which he called an excentrical elliptoid.

Most likely the situation was as follows. Hooke, not having the necessary mathematical technique, was unable to solve exactly the equations of motion obtained from the inverse square law and, in order to find the orbits, he integrated these equations numerically, graphically or on an analogue machine like the concave surface he mentioned. It is known that Hooke had such a machine: he investigated the nature of motion under various laws of attraction, modelling the attraction by the action of a surface on a weight sliding over it. (We observe that all this happened six years before Newton wrote his book and stated the general laws of mechanics. Accordings to our modern ideas, at that time there was no mechanics. Nevertheless, in these pre-mechanics times Hooke found approximate solutions of the equations of motion under the inverse square law, and Huygens stated the law of conservation of energy. It is true that Huygens did not give it in its most general form, but in his formulation (7) the law was applicable in our case, and made it possible to realize that in the absence of air resistance the orbits of a stone inside the Earth must be closed.) Having integrated the equations of motion, Hooke drew the orbits and saw that they were similar to ellipses. This is how the word elliptoid arose. His scientific honesty did not allow him to call them ellipses, since he could not prove that they were elliptic. Hooke suggested to Newton that he do this, saying that he did not doubt that Newton with his superior methods could cope with this problem and also check that Kepler's first law (which asserts that the planets move in ellipses) also follows from the inverse square law.

In sending Newton a letter with this suggestion, Hooke was turning to later discoveries, since he had no time for the mathematical details. Newton was silent and never wrote any more to Hooke (except for one case, when he sent Hooke a request from an Italian doctor who wished to collaborate with the Royal Society and took the opportunity to thank him for his information about his experiments with falling balls), he never referred to the correspondence (although he kept the letters) and he did not speak about the fact that Hooke had posed him the problem of gravitation.

But Newton took up this problem, investigated the law of motion, checked that elliptic orbits had actually been obtained, and proved conversely that the inverse square law follows from Kepler's law on ellipticity of orbits*. In order to put this properly into shape and present it in accessible form, he needed to state the basic principles, referring to such general concepts as mass, force, acceleration. This is how the famous "three laws of Newton" appeared, to which Newton himself did not pretend (the first law, Galileo's law of inertia, had been well known for a long time, and the other two could not have been discovered later than, say, Hooke's law of elasticity or Huygens' formula for centrifugal force). But in connection with the law of universal gravitation Newton behaved less scrupulously.

§4. The *Principia*

On the initiative of the astronomer Halley (1656–1742) Newton wrote a paper with a detailed presentation of his results under the title "Philosophiae Naturalis Principia Mathematica" and sent it to the Royal Society on 28 April 1686. In the manuscript Hooke

---

*   The proofs are given below, on pp. 95–96.

was not mentioned once. Halley, who was a friend of them both, was not at all pleased at this, and he persuaded Newton to insert a reference to Hooke. Newton yielded to persuasion, but did it in a very original form. He wrote that the inverse square law corresponds to Kepler's third law "as Wren, Hooke and Halley independently asserted". Wren and Halley were not, of course, random people. Wren was an architect, one of the founders of the Royal Society, who together with Hooke was occupied with the rebuilding of London after the Great Fire of 1666, and took an active part in the discussion on questions of the motion of bodies. Halley, who subsequently predicted the return of the comet bearing his name, put much effort into forcing Newton to write this book, and his experiments with clocks on the island of St. Helena served as experimental confirmation of the law of gravity. So, putting Hooke between them, Newton not only belittled his role, but also deprived him of the support of friends in the argument about priority which was soon to start.

Here it is appropriate to say a few words about the material position of our heroes. Hooke was poor and lived on the salary that the Royal Society paid him. He did some additional work, using his extensive knowledge of mechanics in carrying out the huge restoration work after the Fire of London. These architectural earnings helped him in the end to create a certain prosperity. Holding the chair at Cambridge, Newton earned considerably more (200 pounds a year), and the farm that he had inherited, which he leased out and where the famous apple tree grew, gave him roughly the same income. Despite the fact that Newton was quite well off, he did not want to spend any money on the publication of the book, so he sent the *Principia* to the Royal Society, which decided to publish the book at its own expense. But the Society had no money, so the manuscript lay there until Halley (who was the son of a rich soap manufacturer) published it on his own account. Halley took on himself all the trouble of publishing the book, and even read the proofs himself. Newton, in correspondence at this time, called it "Your book"...

In this correspondence with Halley, Newton, answering the request to refer to Hooke, wrote a remarkable phrase, which revealed his opinion on the difference between mathematicians and physicists. Newton regarded himself as a mathematician and Hooke as a physicist. Here is how he describes the difference in the approaches of a mathematician and a physicist to natural science.

"Mathematicians, that find out, settle and do all the business, must content themselves with being nothing but dry calculators and drudges, and another that does nothing but pretend and grasp at all the things must carry away all the inventions as well of those that were to follow him as of those that went before... And... I must now acknowledge in print I had all from him and so did nothing myself but drudge in calculating, demonstrating and writing upon the inventions of the great man".

It must be said that Newton made all the discoveries contained in the *Principia* without using analysis, although he had command of it at this time. He proved everything that was required by means of direct elementary geometrical arguments more or less equivalent to analysis (and not by translating analytical calculations into geometrical language).

§5. Attraction of spheres

Let us examine, as an example of Newton's arguments, how he proved that external layers do not act on a stone inside the Earth, that is, that *the field of gravitation inside a homogeneous sphere is zero*. Earlier this fact was studied at (Russian) high school, but it has now fallen out of the programme; hence possibly this remarkable proof is not universally known.

26

Let us consider a point P inside a ball bounded by an infinitely thin spherical layer (Fig.6), take a small solid angle with vertex at P, and prove that the forces, with which two infinitesimal volumes cut out by this angle from the spherical layer act on a body placed at P, balance. (Nowadays, when teaching analysis, it is not very popular to talk about infinitesimal quantities. Consequently present-day students are not fully in command of this language. Nevertheless, it is still necessary to have command of it.) These two volumes are infinitesimal prisms (their generators, of course, slightly diverge, but the amount of this opening can be neglected, since the error is an infinitesimal of higher order), whose volumes can be calculated from the formula $V = lS$, where l is the length of the slant edge, and S is the area of the cross-section. But the edges of our prisms are equal as segments cut out on a line by a pair of concentric circles, and the cross-sections are related like the squares of the distances from P. Thus, these two volumes attract the body at P on opposite sides with identical forces. In exactly the same way, all the other influences balance, so the resultant of all the forces is zero.

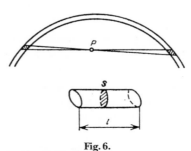

Fig. 6.
Attraction of a spherical layer according to Newton

This sample of Newton's argument shows how it is possible to solve problems from potential theory without analysis, without knowing the theory of harmonic functions, or the fundamental solution of the Laplace equation, or the simple and double layer potentials. Similar arguments, preceding the rise of analysis, often occur in papers of those times and turn out to be very powerful. Here is an example of a problem that

27

people like Barrow, Newton and Huygens would have solved in a few minutes *(8)* and which present-day mathematicians are not, in my opinion, capable of solving quickly*: to calculate

$$\lim_{x \to 0} \frac{\sin \tan x - \tan \sin x}{\arcsin \arctan x - \arctan \arcsin x}.$$

Newton also proved that a homogeneous ball (or a spherical layer) attracts points of the outer domain as if all its mass were concentrated at the centre. Newton's proof is elementary, but not easy (just as it is not easy to calculate the corresponding integral). The modern proof given below (due to Laplace) is unfortunately beyond the bounds of the science taught in Russian high school (which is not surprising), or even in the Faculty of Mathematics and Mechanics at Moscow State University.

Let us consider the velocity field of an incompressible fluid filling the whole of space and spreading spherically-symmetrically over the radii from a source at the origin. The speed of such a flow is inversely proportional to the square of the distance from the source.

Indeed, because of the incompressibility, through each sphere with centre at the source there passes in unit time the same flow of the fluid (as much as the source produces). Because of the spherical symmetry of the current, this flow is equal to the product of the area of the sphere and the speed of flow through it. But the area of a sphere is directly proportional to the square of the radius. Hence, in order that the flow should not depend on the radius, the speed must be inversely proportional to the square of the distance from the source.

Thus, the law of decrease of the rate of a spherically symmetric flow of an incompressible fluid with the distance from the centre is the same as the law of decrease of the force of gravity. (Hence

---

* The only exception I know – G. Faltings – proves the rule.

it is obvious what is the natural analogue of the gravity field in n-dimensional space: the force must decrease in inverse proportion to the $(n - 1$st power of the distance.)

We have proved that the field of the force of attraction of a material point has the following remarkable property of incompressibility: if we regard it as the velocity field of a current, then the value of the flow through the boundary of any bounded domain not containing the attracting point is zero: as it flows in, so it flows out.

It turns out that under any mass distribution the field of the force of attraction by these masses outside these masses has the same property of incompressibility. For under addition of the velocity fields the magnitudes of their flows through any surface are added. Therefore under addition of the velocity fields of two flows of an incompressible fluid we again obtain the velocity field of an incompressible fluid: the flow of the total field through the boundary of the domain is zero if the flows of the fields to be added are zero. Thus, the total force of attraction by several masses has the property of incompressibility (in a domain outside the attracting masses).

In particular, let us consider the field of the force of attraction by a homogeneous ball (or a spherical layer). In the outer domain this field coincides with the velocity field of an incompressible fluid (as we have just proved). It is spherically symmetric. But the only spherically symmetric velocity field of an incompressible fluid is inversely proportional to the square of the distance from the centre. Hence the ball (or layer) attracts external points just as a mass placed at the centre. That the mass at the centre must coincide with the total mass of the ball (or layer) is obvious by comparing the flows through the sphere that envelops the ball in question.

The theorem that a layer does not attract interior points also follows from this argument.* [Both theorems of Newton (on the

---

* PROBLEM. Calculate the mean value of the function $1/r$ over the sphere $(x - a)^2 + (y - b)^2 + (z - c)^2 = R^2$ and of the function $\log 1/r$ over the circle $(x - a)^2 + (y - b)^2 = R^2$.

attraction by spherical layers of interior and exterior points) can be extended to the attraction by layers between homothetic ellipsoids (the role of the centre is played here by any confocal ellipsoid smaller than the one in question). The ellipsoids can even be replaced by algebraic surfaces of any degree *(9)*. The only important thing is that the surface should be hyperbolic (it should intersect every line emanating from some point as many times as the degree of the equation of the surface).]

§6. Did Newton prove that orbits are elliptic?

To conclude the account of the law of universal gravitation we must say a few words about the discussion that has developed around it very recently in the physics journals. In the past this discussion would have been impossible, but now the situation has changed thanks to the fact that the spirit of modern mathematics has penetrated to a number of physicists, causing them, as we shall now see, some damage. They have begun to worry about questions that earlier nobody would have talked about seriously. Many physicists have taken part in this discussion (an account of it can be found in an article by R. Weinstock *(10)*, and the theme of the argument has been stated in the following way: did Newton prove that Kepler's first law follows from the law of universal gravitation?

In reality, this is the question. For the trajectory of a body moving under the action of the force of gravity Newton's laws give the differential equation

$$\ddot{\vec{r}} = -\frac{k\vec{r}}{r^3}.$$

Instead of solving it according to all the rules of science, Newton in his book presented many solutions of this equation and verified that for any initial condition there is a solution among them that satisfies it. In other words, for any point and vector in space in the set of orbits found by Newton there is one that initially passes through this point and has a given velocity vector there. If the initial speed of the body is not too large, then the orbit is elliptic. But who said, ask the physicists experienced in the mathematical niceties, that there does not exist any other trajectory satisfying the same initial conditions along which the body can move, observing the law of universal gravitation, but in a completely different way? Mathematicians know that the absence of another trajectory of this kind is called the uniqueness theorem. Thus, in order to deduce from the law of universal gravitation that a body moves in this and no other way, Newton needed not only to produce many solutions of the differential equation but also to prove the uniqueness theorem for it. Did he do this? No. Well then, generally speaking, it is impossible to use this law to describe reality until the uniqueness theorem has been proved. Who was the first to do this? Johann Bernoulli. So it was he, not Newton, who derived Kepler's law from the law of universal gravitation, and all the glory must belong to him. Here is what the physicists who took part in the discussion say, repeating what was said many years ago by mathematicians (for example, in the book of A. Wintner in 1941).

In fact, all this argument is based on a profound delusion. Modern mathematicians actually distinguish existence theorems and uniqueness theorems for differential equations and even give examples of equations for which the existence theorem is satisfied but the uniqueness theorem is not *(11)*. So various troubles can arise, and if Newton's equation were troublesome, it would actually be impossible to make any deductions. A mistaken point of view arises because of the unwarranted extension of the class of functions under consideration. The fact is that in modern mathematics the concepts of function, vector field, differential equation have acquired a different meaning in comparison with classical mathematics. Speaking of a function, we can have in mind a rather nasty object – something differentiable once or even not at all – and we must think about the function class containing it, and so on. But at the time of Newton the word function meant only very good things. Sometimes they were polynomials, sometimes rational functions, but in any case they were all analytic in their domain of definition and could be expanded in Taylor series. In this case the uniqueness theorem is no problem, and at that time nobody gave it a thought.

But in reality Newton proved everything, to a higher standard. The following theorem is true.

Suppose we have a differential equation

$$\dot{x} = v\,(t,\,x)$$

and that for any initial condition a we have produced a solution $x\,(t,\,a)$ with $x\,(0,\,a) = a$, where this solution depends smoothly (that is, infinitely differentiably) on a. Then the uniqueness theorem is true for this equation.

This theorem can be proved very easily. From the existence of a solution that depends smoothly on the initial data it follows that there is a (local) diffeomorphism that rectifies the original field of directions, taking it to the standard form of a field of horizontal directions (our solution gives this diffeomorphism: $(t,\,a) \leftarrow (t,\,x(t,a))$). The uniqueness theorem is obviously satisfied for the rectified field, since the equation takes the form $\dot{a} = 0$.

Thus, in general, uniqueness does not follow from the existence of a solution, but everything will be in order if the solution produced depends smoothly on the initial condition.

Let us see what Newton did. For each initial condition he produced a solution, described it, and from this description it became obvious straight away that the solution depends smoothly on the initial condition. Thus, there is no doubt about the uniqueness and Newton correctly proved Kepler's first law.

Of course, one could raise the objection that Newton did not know this theorem. In fact, he did not state it in the form that we have just used. But he certainly knew it in essence, as well as many other applications of the theory of perturbations – the mathematical analysis of Newton is to a considerable extent a well developed theory of perturbations.

CHAPTER 2.
MATHEMATICAL ANALYSIS

§7. Analysis by means of power series

Newton remarked that the laws of nature are expressed by the differential equations that he devised. Individual, and at times very important, differential equations had been considered and solved even earlier, but Newton turned them into an independent and very powerful mathematical instrument.

Newton discovered a way of solving any equations, not only differential but, for example, algebraic. He regarded this discovery as his most important achievement and codified it in a letter to Leibniz on 24 October 1676 (it was sent via Oldenburg and has therefore gone down into history as the "Second letter to Oldenburg" (epistola posterior)), in which he described analysis.

Analysis is a concept that is quite difficult to define. Newton understood by analysis the investigation of equations by means of infinite series. In other words, Newton's basic discovery was that everything had to be expanded in infinite series.* Therefore, when he had to solve an equation, whether a differential equation or, say, a relation defining some unknown function (this is now known as one form of the implicit function theorem), Newton proceeded as follows. All functions are expanded in power series, the series are substituted into one another, the coefficients of identical powers are compared, and one by one the coeffi-

---

* "These studies [on power series] stand in the same relation to algebra as the studies of decimal fractions to ordinary arithmetic" *(12)*.

cients of the unknown function are found. The theorem about the existence and uniqueness of solutions of differential equations is proved in this way instantaneously together with the theorem about dependence on the initial conditions so long as we are not worried about the convergence of the resulting series. As for the convergence, these series converge so rapidly that Newton, although he did not strictly prove convergence, had no doubts about it. He had the definition of convergence and explicitly calculated series for specific examples with an enormous number of digits (in the letter to Leibniz Newton wrote that he was ashamed to admit to how many digits he took these calculations). He remarked that his series converge like a geometric progression and so there were no doubts about the convergence of his series. Following his teacher Barrow, Newton realized that analysis has a justification, but quite reasonably he did not think it useful to linger on it ("One could extend it by an apagogical* argument", wrote Barrow, "but to what purpose?").

§8. The Newton polygon

Apart from power series in which solutions of differential equations are expanded, Newton also used fractional powers, which are used when one needs to find an expansion for an algebraic function $y(x)$ defined by an equation $f(x, y) = 0$. Suppose, for example, that we need to solve the algebraic equation

$$ax^2 + by^3 + cxy^2 + dx^7 = 0.$$

Then, says Newton, we need to make the following transformation (similar transformations are now called Fourier transformations, but in the given case they are nevertheless *Newton transfor-*

---

* "By contradiction", that is, strict.

mations). We cease to regard a polynomial as a function of the variables x and y, and consider it as a function on the integer lattice in the plane. At a point with coordinates (m, n) this function takes a value equal to the coefficient of $x^m y^n$ in the polynomial. We now mark on the lattice the points corresponding to monomials with non-zero coefficients, and take their convex hull. We obtain the Newton polygon (Fig. 7). It turns out that the monomials corresponding to vertices inside the polygon have no influence on the form of the series, so we can forget about them, and we need only consider the sides. For example, the side AB in Fig. 7 coresponds to the two-term equation $ax^2 + by^3 = 0$. Solving this equation, neglecting everything else, we find y as a function of x: $y = kx^{2/3}$. This function gives a good approximation near the origin to the solution of our equation. If we wish to find the next approximation, we need to write $y = kx^{2/3} + z$ and substitute it into the original equation. After the substitution we again obtain an algebraic equation, but now for z, which we need to deal with in exactly the same way. Iterating this process, we obtain a series in fractional powers (it is now called the Puiseux series), which gives the solution $y(x)$ of the equation in a neighbourhood of the origin. This method always works. If we had started with another side of the polygon, we would have arrived at another series, which corre-

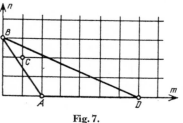

Fig. 7.
The Newton polygon

sponds to another branch of the algebraic function. The side BD in Fig. 7 corresponds to the asymptotic behaviour at infinity.

Here I have given an account of this small part of Newton's work, of which he was very proud and which is contained in the same letter to Oldenburg in 1676, partly because unfortunately it is not mentioned to students in the first course, although it is a basic working apparatus in local analysis, and it is also very beautiful. In modern mathematics Newton polygons (and polyhedra) occur in the geometry of toric manifolds, which arose about 1973*, and also in physics and mechanics in the theory of similarity, dimension, and scaling.

The letter to Oldenburg was intended for Leibniz, as we have already mentioned. But Leibniz lived in Germany and Newton in England, and at that time it was not safe to correspond with foreign scholars. Newton did not send the letter to Leibniz, but directed it to the secretary of the Royal Society Oldenburg, so that the letter would go by the official route. Oldenburg passed this letter to Leibniz. Newton's precaution was not superfluous. The too sociable Oldenburg spent some months in the Tower for his contact with foreigners.

§9. Barrow

We now consider the beginnings of analysis, and I start with an account of Barrow. Newton's teacher Isaac Barrow was born in 1630 and died in 1677 *(13)*. In contrast to the shy and bashful Newton, who even when elected as representative for Cambridge in Parliament did not utter a word there (true, it is said that Newton once got to his feet at a session of Parliament, but with a very brief speech: he suggested closing the window), Barrow in his youth was a very wild person. His father – a London linen merchant – made sure that his son could not enter the commercial world because of the wildness of his character, and sent him off to study.

Barrow learnt various subjects, but he was most attracted by theology. The way of thinking that determined his future path was this: in order to be a good theologian, it is necessary to know chronology...

---

* A.G. Khovanskii, The geometry of formulas, Soviet Scientific Reviews – C, Mathematical Physics Reviews, Vol. 4, Harwood, New York, 1984, pp. 67–90.

38

The idea that chronology is a very important science was obvious to everybody at that time, including Newton. And at present some mathematicians, probably following Barrow and Newton, though not in England, but in Moscow, are keenly interested in problems of chronology *(14)*. Newton was very seriously engaged in the chronology of ancient Egypt. The following problem arose in it. So much historical information, discovered up to this time, had already accumulated that did not agree with the biblical dates of the creation of the world. According to the Bible, the duration of Man's existence, from Noah to the birth of Christ, was 2348 years, but there were many Pharaohs and dynasties, and there was not room for all of them. Newton wrote special texts in which he suggested a way out of this difficulty. He found Pharaohs in the bible whose name began with the letter S (Sheshonk, or Shishak) and Herodotus mentioned another Pharaoh with a different name, but also beginning with S (Sesostris, now called Senurset). Newton suggested that these two Pharaohs should be regarded as one, accordingly correcting the ancient Egyptian chronology (shortening it by 2000 years – entirely in the spirit of modern mathematicians). But apparently more scientific approaches to this question were maturing at this time. Barrow, for example, suggested using information about eclipses. He therefore said: "To be a good theologian it is necessary to know chronology, which requires a knowledge of astronomy".

Familiarity with astronomy in turn led Barrow to geometry. This happened for two reasons. Firstly, it was necessary to point a telescope at the stars, and secondly it was necessary to make the telescope work, that is, to grind lenses for it. Newton's first activity was connected with astronomy. He made the first reflector – a metal telescope *(15)*.

Barrow began by discovering his remarkable lens formula $1/f = 1/r_1 + 1/r_2$, which is now taught in Russian high school (without mentioning Barrow by name). In order to derive this formula one needs to be able to draw tangents to the lens, to find the points of intersection of infinitely close normals, and to analyse the resulting focal points. Thus Barrow was interested in geometry.

It must be said that his life was very hard, because in England at that time the survivals of feudalism were very powerful, dynasties were changing, revolutions were occurring, people were subject to persecution for their religious beliefs (at that time this was regarded as quite permissible, even in the most civilized countries). Barrow's religious ideas did not coincide with those that prevailed in England at that period. He therefore needed to go away somewhere, and he set off for the Holy Land to get everything in place. But during the voyage pirates attacked the ship, and although Barrow, the only passenger with a sword in his hands, took part in the boarding battle and overcame the pirates, he did not reach the end of the voyage. Barrow safely returned to England, where at that time the dynasty and the religious situation had changed. He was therefore able to obtain the chair of mathematics founded by Henry Lucas, who had left money to Trinity College, Cambridge. Barrow, who was studying geometry at this time, began to give lectures in mathematics there. Newton was then a second-year student, and apparently attended these lectures. Barrow's lectures were subsequently published, and Leibniz bought one of his books in 1673. True, Leibniz later said that he rarely saw a man or a book of which he could not make some use, but he put Barrow's book on a shelf and did not read it.

What did Barrow's lectures contain? Bourbaki writes with some scorn that in his book in a hundred pages of the text there are about 180 drawings *(16)*. (Concerning Bourbaki's books it can be said that in a thousand pages there is not one drawing, and it is not at all clear which is worse.) Because of the abundance of these drawings, in Bourbaki's opinion, nobody has noticed what this book contains, since all its contents are found in geometrical abundance. In this book no new terms or ideas are introduced, and there are neither functions nor derivatives. It was mainly devoted to a development of a single principle, from which many consequences were derived. This principle is that there is duality between problems about tangents and problems about areas.*

This duality enables us, whenever some problem about tangents has been solved (we know how to draw a tangent to a curve or to calculate the subtangent, the normal, and so on), to solve the corresponding problem about areas to which this curve is the answer. In the dual problem it is a question of the area under another curve, which can be obtained geometrically from the first. Thus, this book is actually devoted to the Newton – Leibniz formula, which Barrow could not have known, since this happened twenty years before their first publications. Barrow also derived consequences from this principle. Some of these went quite far. If we look attentively, we can discover two main consequences of this kind: on the one hand, a formula for change of variables in a definite integral, on the other hand integration of ordinary differential equations with separated variables. In the end, coming to the theorem on change of variable in an integral, Barrow added that, unfortunately, he had discovered this fact very late, and if he had known it earlier, much of the previous text could have been simplified. But because Barrow had many other things to do, he did not make these changes in the text.

It is actually very difficult to read Barrow's book *(17)*, and not without reason Bourbaki says that without analysing these 180 drawings it is impossible to understand anything in it. But Newton attended Barrow's lectures and so he comprehended everything and perfectly understood the contents of the book. In many cases he even simplified and improved the presentation, about which Barrow, an extremely conscientious man, informed the reader in a suitable reference. But this help of Newton did not touch at all on the main points – the Newton-Leibniz formula and the solu-

---

* From the papers of Newton that have been preserved it is obvious that he already knew about this duality in 1665 or 1666, possibly independently of Barrow.

tion of equations with separated variables. This was Barrow's own contribution, and Newton never pretended to these discoveries.

Barrow, having noticed such a talented pupil, who in his 27 years had already made some deep discoveries in improvements to the telescope, in optics, and in geometry, thought that he was already too old to take up his chair (he was 39) and had decided to go over to ideological work, so he handed over the chair to Newton. Later he ran into great difficulty. The fact is that Barrow was in holy orders and only because of this was he able to keep holding the chair. In those far-off times it was impossible to keep holding the chair without taking holy vows. Newton did not wish to take vows (although he was promised the deanship). So Newton could not remain in the chair at Cambridge for more than seven years. But Barrow was an influential man and, on leaving Cambridge, became a court preacher in London. He was therefore able to obtain special permission from the king for Newton as an exception. So Newton kept his chair at Cambridge and carried out very fruitful activity there.

§10. Taylor series

Integration had already been encountered with Archimedes, and differentiation with Pascal and Fermat; the connection between these operations was known to Barrow. What did Newton do in analysis? What was his main mathematical discovery? Newton invented Taylor series, the main instrument of analysis.

Of course, some perplexity may arise here, connected with the fact that Taylor was a pupil of Newton and his corresponding paper dates from 1715. We can even say there are no Taylor series at all in Newton's work. This is true, but only partly. Here is what actually happened. Firstly, Newton found the expansions of all the elementary functions – sine, exponential, logarithm, and so on – in Taylor series and thus verified that all the functions that

42

occur in analysis can be expanded in power series. He wrote out these series – one of them is called Newton's binomial formula (of course, the exponent in this formula does not have to be a natural number) – and used them constantly. Newton correctly assumed that all the calculations in analysis need to be carried out not by repeated differentiations, but by means of expansions in power series. (For example, he used Taylor's formula for calculating derivatives rather than using the derivatives for the expansion of functions – the last point of view was unfortunately supplanted in the teaching of analysis by the clumsy apparatus of the infinitesimals of Leibniz.) Newton derived a formula analogous to Taylor's series in the calculus of finite differences – Newton's formula – and finally he had Taylor's formula in the general form, only in those places where factorials should be there were coefficients not written out explicitly.

Newton could have said what the coefficients should be there* (he put factorials in the formula for finite differences), but he did not think it was necessary to do this. He probably had a psychological reason for this. The fact is that, for Newton, quantities were not abstract numbers, they had a certain physical existence. But all the quantities h^n in Taylor's formula have different dimensions, and for everything to be in order there should be a coefficient with the appropriate dimension before each of them. But then in a different system of units the coefficient would also be different. There was no unique system of units at that time; the units of measurement changed from country to country and even from county to county. Therefore people preferred not to state the dimensional coefficients in the statement of laws, but to speak merely of proportionality, as, for example, in Hooke's law: "the extension is proportional to the force". If Taylor's formula is also written in such a dimension-free form, then the factorials disappear and we find

---

* In Newton's papers that have been preserved the series was written out completely (I am grateful to A.P. Yushkevich for this information).

that the terms in the increment of a function are directly proportional to the n-th derivative and the n-th power of the increment of the argument. After this each coefficient of proportionality can be found, depending on the units that are being used.

In Newton's formula in the calculus of finite differences the factorial coefficients are written out explicitly, since in this case the unit of measurement is fixed by the choice of step of the lattice. The clumsy notation for higher derivatives that Leibniz used is convenient here because it automatically takes account of the dimensions, so the formulae appear the same for any system of units.

Newton did not publish his discoveries in the field of analysis. He merely informed Leibniz that he was able to "compare the areas of any figures in half of a quarter of an hour". I do not know whether this level of mastery of analysis is attained by present-day first-year students.

§11. Leibniz

When speaking of the history of analysis, it is impossible not to say a few words about the rivalry between Newton and Leibniz. Newton was very seriously concerned with questions of priority. Somewhat earlier he had stated the following principle: each person must one day make a choice – either to publish nothing, or to devote all his life to the struggle for priority. For himself Newton apparently also made his decision on this, choosing both policies together: he published hardly anything, and he was constantly struggling for priority.

As for the invention of analysis, here the first publications were due to Leibniz, who said that he developed his differential and integral calculus independently of Barrow and Newton. Nevertheless, discussion on this point flared up so fiercely that as a result the opinion grew that it would be better not to contest priority than to carry on such discussion.

Gottfried Wilhelm Leibniz (1646–1716) was a diplomat of the Elector (Kurfürst) of Mainz*, sent by him in 1672 to Paris in very difficult conditions.

At that time France was already a united absolute power under the authority of Louis XIV, who was very powerful militarily, but Germany was fragmented and could not in any way oppose the military might of Louis XIV, whose cavalry could cover the whole of Germany in the course of a day. The Germans were very much afraid of this and wished to find some way out. Leibniz, with his inherent diplomatic ability, devised a method of rescuing Germany from French invasion and was sent to Paris to carry out this plan. Leibniz's method was as follows: he wished to palm off on Louis XIV the project of conquering Egypt. Leibniz formed an appropriate project and actually delivered it to the French government. The French government carried out the project, but not for some time, in fact only under Napoleon, but the idea goes back to Leibniz.

Although Louis XIV did not carry out Leibniz's project, the visit to Paris was not in vain. Leibniz made the acquaintance of Huygens there. Huygens was a Dutch scientist, but in 1666 he was invited to France to be the first chairman of the Academy. Later, after the abolition of the Edict of Nantes and persecution for religious beliefs had been restored in France, he decided to return to Holland.

From Huygens Leibniz learnt about the existence of some very interesting mathematical papers. Leibniz had been interested in mathematics earlier, because he was always interested in anything general and had all sorts of general ideas.**

---

* A peace movement founder and a liberal ruler, who even abolished the burning of witches in his principality.

** Among other things, Leibniz was responsible for the idea and project of the Russian Academy of Sciences, formed on the instructions of Peter I, and – also on the instructions of Peter I – the project for a reconstruction (perestroika!) of the Russian legal system, having in mind the transformation of this country into a law-based state – a goal still to be achieved.

For example, he considered it necessary to unite all the religions, or if not all, then at least all the Christian religions, or if the orthodox Christians did not agree, then at least the Catholics and Protestants, and if this was impossible, then to unite at least all the Protestants. True, this was not successful, though he applied all his forces.

In exactly the same way, Leibniz considered it necessary to discover the so-called characteristic*, something universal, that unites everything in science and contains all answers to all questions. He also devised all possible universal methods for solving all problems straight away**. For example, he manufactured calculating machines following Pascal. (In contrast to the arithmometer of Pascal, the arithmometer of Leibniz made it possible to take square roots.) The arithmometer itself was not preserved, but evidence has come down to us of a visit by Leibniz to England, where he demonstrated his work to British scientists (Hooke immediately improved his construction).

Thus, Leibniz liked mathematics very much, he wished to unite all its methods, and Huygens advised him to study Pascal. From Pascal's successors Leibniz got hold of his letters and notes (later lost) and found in the papers of Pascal a picture which represented the celebrated differential triangle (Fig. 8).

---

*   A. N. Parshin explained to me that Leibniz's "characteristic" essentially coincides with the "Gödel numbering", by means of which Gödel proved the incompleteness of all sufficiently rich theories, thus disproving the Leibniz-Hilbert programme of formalizing mathematics.

**   "A good legacy is better than the most beautiful problem of geometry", wrote Leibniz to l'Hôpital, "since it plays the role of a general method and enables us to solve many problems". *(18)*
Reference to the idea of universality does not justify the cynicism of this joke of Leibniz: a similar blasphemous phrase would have been unthinkable in the mouth of Barrow and even Newton.

At that time Descartes, Fermat and Pascal were able to differentiate polynomials and knew how to draw tangents to parabolas of all degrees, and the fundamental infinitesimal triangle was already explicitly present in Pascal's work. In the work of other geometers – Huygens and Barrow – many objects connected with a given curve also appeared. In Fig. 8, for example, there are the following quantities: abscissa, ordinate, tangent (the segment of the tangent from the abscissa axis to the point of contact), the slope of the tangent, the area of a curvilinear figure, the subtangent, the normal, the subnormal, and so on. Usually all these objects are considered separately. Barrow, for example, derived relations between the subnormal and the subtangent by means of a new curve, drawn in a new plane. Leibniz, with his individual tendency to universal-

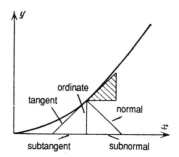

Fig. 8.
The infinitesimal Pascal triangle and various functions

ity, decided that all these quantities should be considered in the same way. For this he introduced a single term for any of the quantities connected with a given curve and fulfilling some function in relation to the given curve – the term *function*. Examples of functions were all the quantities that occur in Fig. 8, for example, abscissa, ordinate, subnormal, subtangent, and so on.

Thus, according to Leibniz many functions were associated with a curve. Newton had another term – fluent – which denoted a flowing quantity, a variable quantity, and hence associated with motion. On the basis of Pascal's studies and his own arguments Leibniz quite rapidly developed formal analysis in the form in which we now know it. That is, in a form specially suitable to teach analysis by people who do not understand it to people who will never understand it. Leibniz wrote: "A poor head, having subsidiary advantages,... can beat the best, just as a child can draw a line with a ruler better than the greatest master by hand" (Guhrauer, Leibniz's deutsche Schriften, vol. I, pp. 377–381). Leibniz quite rapidly established the formal rules for operating with infinitesimals, whose meaning is obscure.

Leibniz's method was as follows. He assumed that the whole of mathematics, like the whole of science, is found inside us, and by means of philosophy alone we can hit upon everything if we attentively take heed of processes that occur inside our mind. By this method he discovered various laws and sometimes very successfully. For example, he discovered that $d(x + y) = dx + dy$, and this remarkable discovery immediately forced him to think about what the differential of a product is. In accordance with the universality of his thoughts he rapidly came to the conclusion that differentiation is a ring homomorphism, that is, that the formula $d(xy) = dxdy$ must hold. But after some time he verified that this leads to some unpleasant consequences, and found the correct formula $d(xy) = xdy + ydx$, which is now called Leibniz's rule. None of the inductively thinking mathematicians – neither Barrow nor Newton, who as a consequence was called an empirical ass in the Marxist literature – could ever get Leibniz's original hypothesis into his head, since to such a person it was quite obvious what the differential of a product is, from a simple drawing (Fig. 9). Clearly, the increment of the area of a rectangle consists of three terms: the areas of the two infinitely thin rectangles xdy and ydx, and an infinitesimal of higher order $dxdy$, which can be neglected. Having such a geometrical interpretation before

48

his eyes, he would never suspect that the required increment was equal to this neglectable quantity. But for the scholastic Leibniz such an algebraic way of thinking was very typical.*

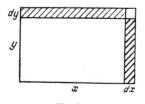

Fig. 9.
Leibniz's formula

§12. Discussion on the invention of analysis

It is necessary to say a few words about the ugly dispute that flared up between Newton and Leibniz after Leibniz published his infinitesimal calculus in 1684.

For about ten years everything was quiet and peaceful, but then the pupils of Newton and Leibniz began to argue about which of the two first invented analysis.

The term *analysis* was used by Newton in the sense of "investigation" (of curves by means of power series). Newton regarded analysis as a development of Descartes' analytical geometry, which he valued highly.

For authors of high school programmes and textbooks it may be of interest that Newton studied the elements of geometry not according to Euclid, as was usual (and is still usual essentially), but according to Descartes, and that he invented analysis possibly because of this unusual way of studying geometry.

Later, on the advice of Barrow, who discovered his pupil's unfamiliarity with Euclid in an examination, Newton carefully worked through Euclid and virtuously mastered the technique of

---

*   As late as 1686 Leibniz assumed that the circle of curvature intersects a curve in no fewer than four coincident points.

the ancients, but initially he did not like Euclid, because he thought it was foolish to prove things that were quite obvious.

Newton's analysis was the application of power series to the study of motion, that is, functions and mappings, as we would now say. For Leibniz, as we have seen, analysis was a more formal algebraic study of differential rings.

Here are some details of this dispute, which show clearly that one should never engage in disputes of this kind, because such remarkable people and great mathematicians as Newton, Leibniz, and Johann Bernoulli appear here in a terrible light.

In this episode, for example, there appears an anonymous letter. This letter, published by Leibniz unsigned in the "Journal des Savants", was written by Johann Bernoulli, a pupil of Leibniz. In it he defends Leibniz and attacks Newton. But a year later Newton and his pupils, answering this letter, called it Bernoulli's letter. Bernoulli, seeing that he had not been able to preserve his incognito, reproached Leibniz in that this had caused him to fall out with Newton, who had just introduced Bernoulli to the Royal Society, and had promised to elect his son in the future. "I was astonished", wrote Johann Bernoulli to Leibniz, "how Newton could have known that I wrote the letter, since nobody knew, except you, to whom I sent it, and I, who wrote it". But some time later he was able to unravel the mystery. In one place there crept into the letter the expression "meam formulam", that is, a reference to "my formula". Since this formula was due to Johann Bernoulli, Newton easily recognized the author of the letter.

Leibniz wrote a letter to Princess Caroline, the wife of the Prince of Wales, warning her that she "should not allow the anti-religious Newton to disturb her simple-minded German faith". At that time (it was about 1715) this was a dangerous accusation for Newton, who then held a high state office – he was Master of the Mint. A state official accused of antireligion could pay dearly for it. Fortunately for Newton, this did not happen in his case *(19)*.

50

Newton also did not behave very well in this episode. He set up a commission whose task was to examine the question of priority and to take the final decision. At this time Newton was President of the Royal Society*, so in the make-up of the commission to give it greater impartiality, in his words, numerous scholars from foreign countries were included.

The commission considered the question of the dispute over priority and published its report. The following words preceded the report of the international authoritative impartial commission: "No one is a proper witness for himself. He would be an iniquitous judge, and would crush under foot the laws of all people, who would admit anyone as a lawfull witness in his own case". It goes on the defend Newton and accuse Leibniz for his unfounded claims of priority concerning the unpublished results of Newton.

Later, after the death of Newton, it became clear from his papers that Newton had guided the drawing up of the report, and the pathetic accusation of an unrighteous judge was written by him personally, and of the "numerous scientists" not from England there were only two, and only one of these was a mathematician.

Leaving this sad episode, from which we should all make deductions about the scientific value of discussions about priority and other matters peripheral to science, I shall say something about those geometrical works that led to the creation of analysis.

---

* Only after the death of Hooke in 1703 did Newton agree to take on the position of President of the Royal Society. One of the first acts of Newton in this position was to destroy all the instruments of the late Hooke, and also his papers and portraits. So now the Royal Society had portraits of all its members except Hooke. Not one drawing of Hooke, who was a member, curator and secretary of the Royal Society, was preserved. In the folder of Hooke's biography recently published in the Soviet Union *(20)* there is a portrait, but this portrait is not genuine, but made up by the methods of modern crime detection from verbal descriptions of Hooke.

CHAPTER 3.
FROM EVOLVENTS TO
QUASICRYSTALS

§13. The evolvents of Huygens

To Newton analysis was necessary as a basis for the investigation of curves, which arise in mechanics and in geometry. We have already seen some ways in which curves arise. Other ways were found by Huygens, who investigated a number of problems in analysis, optics and mechanics. For example, 11 years before the first publications of Leibniz on analysis and 13 years before the appearance of "Newton's laws" Huygens published his calculation of the centrifugal force in motion in a circle (that is, he twice differentiated a vector-valued function and used "Newton's second law").

Huygens solved all the problems by means of elementary geometrical constructions, but he obtained significant results.

One of Huygens' important achievements was the investigation of evolvents, which he introduced. Evolvents occur in many old textbooks on analysis, beginning with the first textbook of l'Hôpital and going roughly up to Goursat, but in modern courses there is a tendency to pass over them.

Suppose we are given a curve. An *evolvent* of it is the trajectory described by the end of a stretched string unwinding from our curve (Fig. 10). A remarkable property of an evolvent is that it has a cusp at the point P and if we attempt to describe it in a neighbourhood of this point by means of the Taylor series we see that it is not smooth there, although it appears to be (and, moreover, it has tangents at all points). The lack of smoothness follows from the fact that the radius of curvature at a point X of the evolvent is equal to the length of the free end of the string YX,

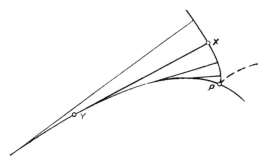

Fig. 10.
Formation of an evolvent by means of a string

and since the string becomes shorter as X approaches P, the curvature at P becomes infinite, and so the point P itself is singular. It turns out that at this point the evolvent has a singularity of type $3/2$, that is, in a neighbourhood of P it is diffeomorphic to the semicubical parabola $y = x^{3/2}$. Why is the evolvent drawn with two branches in the figure? Firstly, if we write the equation of the semicubical parabola in the form $y^2 = x^3$, we see that it has a second branch, and secondly this can be seen if we look at this picture from the point of view of modern geometry.

Suppose that our curve is convex, and that s is the natural parameter along it (that is, the length of the curve), $\vec{r}(s)$ is the radius vector of the point Y of the curve corresponding to the value of the parameter s, and that t is the length of the free part of the string. Then the radius vector of the point X of the evolvent obtained when the string unwinds from the curve at Y is equal to $\vec{r}(s) + \overrightarrow{YX} = \vec{r}(s) + t\vec{r}\,'(s)$, since s is the length along the curve. We thus obtain a mapping F of the plane with coordinates (s, t) to the plane in which our curve lies. It is easy to see that this mapping is

54

smooth, but it is not a diffeomorphism. Simple analysis of the mapping F shows that its image is part of the plane lying on one side of the curve, and all points of the image not belonging to the curve have exactly two inverse images (see Fig. 11), and at each point of the curve there is only one inverse image. Therefore the

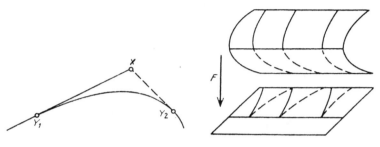

Fig. 11.
Construction of the two inverse images of the point X

Fig. 12.
Formation of evolvents by folding

mapping F is constructed as the mapping of projection onto a plane of the surface folded over our curve (Fig. 12). Such a mapping is called a folding mapping.

It is now obvious how an evolvent arises. This happens if we fix the length of the string, that is, the sum $s + t$. The equation $s + t = a$ defines on the (s, t)-plane a family of parallel lines, which becomes a smooth family of curves on the folded surface and after projection it gives the family of curves below. Each evolvent corresponds to some value of the length a and therefore lies entirely on one of these curves. But each of these curves has two branches. One branch corresponds to the upper part of the curve on the surface, and the other to its lower part. This is how the second branch of the evolvent appears, which corresponds to the lower part. These are two parts of one curve, and the second is an analytic continuation of the first corresponding to negative values of the parameter t (physically this means that the string now winds back on itself; Fig. 13).

The presence of the remarkable singular points on evolvents was discovered by Huygens. He used these remarkable points very well in creating the isochronous pendulum. If a pendulum hang-

Fig. 13.
Formation of the two branches of evolvents by means of the string

Fig. 14.
The isochronous pendulum of Huygens

ing from a string is made to oscillate between sidepieces made in the form of a cycloid (Fig. 14), then it will move along an evolvent of the cycloid (which is also a cycloid) and all its oscillations (that is, not only small, but also large) will have the same period.

§14. The wave fronts of Huygens

Evolvents are connected with an object encountered in another investigation of Huygens, namely in the theory of wave fronts. Huygens, considering the propagation of waves issuing from some source, discovered that singularities can also arise here.

Suppose, for example, that the source has the form of an ellipse and that the waves are propagated inside the ellipse with unit speed. According to Huygens' principle, in time t the wave front will be the envelope of a family of circles of radius t with centres on the ellipse and for small t it will be an equidistant curve of the ellipse. If t increases, then at some instant singularities appear on the envelope (Fig. 15). They will also be of semicubical type. Singular points were very important for Huygens in the investigation of the correspondence between waves and rays, investigations that we now attribute to the calculus of variations, optimization, Hamiltonian mechanics. Therefore in his work in both the theory of waves and the theory of the pendulum, carried out in the 1650s, there were many similar figures with an investigation of all the singularities that arise there.

Let us consider in the plane the region bounded by some curve, and suppose that one of the points of the curve is the source of a perturbation (Fig. 16). Then the fronts consisting of

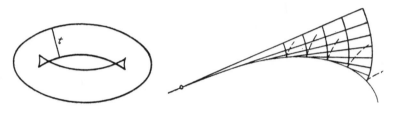

Fig. 15.
Singularities of the wave front

Fig. 16.
Evolvents as wave fronts

points which the perturbation reaches at a definite time, bypassing an obstacle bounded by the curve, will be evolvents of this curve in a neighbourhood of the boundary of the obstacle. Thus, the evolvents of a curve that bounds some region can be regarded as Huygens wave fronts on a manifold with a boundary. Such a front, although it seems at first glance astonishing, has a singularity of type 3/2 at points of the curve (and consequently after analytic continuation a second branch appears on it).

§15. Evolvents and the icosahedron

Further extension of these investigations leads to the study of singularities in three-dimensional space or singularities in the plane, but in a more complex situation. The resulting curves and surfaces also turn out to be very remarkable. Many of them were studied at the time of Huygens or a little later. For example, a famous caustic was introduced by Tschirnhausen soon after Huygens, and the wave fronts and the properties of evolvents connected with them appeared in the first textbook on analysis, written by l'Hôpital from Bernoulli's lectures.

This book examines, in particular, the following case. Suppose we are given a curve in general position in the plane. On such a curve there may be a point of inflexion, but only of the simplest type (the third derivative of the corresponding function is nonzero, since the second and third derivatives cannot vanish simultaneously for a function of general position). We need to explain what the family of evolvents of such a curve looks like. As I understand it, for mathematicians of that time – Barrow, Newton, Leibniz, Bernoulli, and even for his pupil l'Hôpital – this problem was completely within their powers. Of course, it presented certain difficulties for them, but incomparably smaller difficulties than for modern mathematicians. I think that the majority of modern students studying analysis, even the best ones, would not be in a position to construct an evolvent of the cubical parabola $y = x^3$. The answer to this

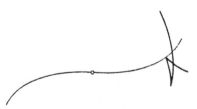

Fig. 17.
An evolvent of a cubical parabola

problem is very remarkable (Fig. 17). As before, the evolvent has a singularity of type $3/2$ on the curve itself, but it also has a singularity of type $5/2$ on the tangent drawn through the point of inflexion. If we change the length of the string, we obtain a family of curves whose singularities fill two curves – the cubical parabola itself and the tangent at the inflexion (Fig. 18). Bennequin (*1*) discovered this picture in l'Hôpital's textbook "Foundations of analysis".

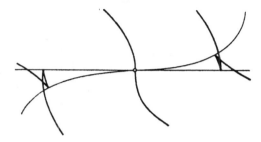

Fig. 18.
The family of evolvents close to the point of inflexion of the curve

Although this picture appears in old works, it became possible to discover it only because it was drawn by modern mathematicians working on another problem, which seems to have no relation to evolvents. In modern mathematics it was discovered that the singularities encountered here are connected with groups generated by reflections. In particular, our picture is connected with the group H_3. This is the group of symmetries of the icosahedron generated by reflections in its 15 planes of symmetry *(21)*.

The appearance of the regular polyhedra is often unexpected.

Kepler, when studying the motion of the planets, stated as well as the three laws that we know a fourth mystical law which states that the major semiaxes of the orbits can be calculated in terms of the regular polyhedra. Ever since then regular polyhedra have turned up equally unexpectedly in other cases where, however, they were more connected with the essence of the matter.

For the symmetry group of the icosahedron we can consider the so-called *discriminant*. Here is how it is obtained. The space $\mathbf{R}^3$

is complexified and turns into $\mathbf{C}^3$, the complex three-dimensional space in which the group H_3 also acts. The quotient space $\mathbf{C}^3$ with respect to this group is again isomorphic to $\mathbf{C}^3$ (this follows immediately from an analogue of a fundamental theorem about symmetric polynomials which, incidentally, has also vanished from the Moscow University algebra course). Thus, here there are three basic invariants in terms of which all the invariant polynomials of this group can be expressed polynomially *(22)*. On the other hand, in the original $\mathbf{C}^3$ there are the mirrors in which the reflections are carried out (15 of them). The number of images of a point not lying on any of the mirrors is equal to the order of the group, that is, 120. For a point on a mirror there are fewer images. A point of the quotient space – an orbit of this group in the original three-dimensional space – is said to be *regular* if the points of which it consists do not lie on the mirrors. The remaining orbits and the points of the quotient space corresponding to them are said to be irregular. The set of all irregular orbits – the image of one mirror under the factorization mapping – is a subvariety of the quotient space. The intersection of this subvariety with the set of real points is a certain variety with singularities, which is called the discriminant of the group H_3.

In exactly the same way, for any other group generated by reflections we can construct a certain variety with singularities.

Here is an example where this can be seen explicitly. In the plane let us consider three lines making angles of 120° with each other (Fig. 19).

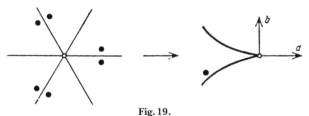

Fig. 19.
A group of reflections, its mirrors, an orbit and the discriminant

60

The group generated by reflections in these lines contains six elements, and a regular orbit consists of six points. These can all be represented as follows. We realize our plane as the plane in three-dimensional space with coordinates z_1, z_2, z_3 given by the equation $z_1 + z_2 + z_3 = 0$. In this three-dimensional space there acts the group of permutations of three elements, which interchanges the coordinate axes. This group is generated by reflections in the mirrors $z_i = z_j$, whose traces on our plane are the original three lines. Therefore the orbits in the plane are simply triples of numbers with zero sum, considered up to permutations. The regular orbits are triples all three of whose numbers are distinct. How can we naturally parametrize these unordered triples of numbers (they are complex, since we have assumed that com-plexification has been carried out), that give zero sum? This method is well known, since an unordered set of three numbers is uniquely defined as the set of roots of a cubic equation $z^3 + \lambda_1 z^2 + \lambda_2 z + \lambda_3 = 0$. But our numbers give a zero sum, so $\lambda_1 = 0$, and the space of orbits of this group generated by reflections is uniquely parametrized by cubic equations of the form $z^3 + az + b = 0$. That is, the quotient space is simply the plane with coordinates (a, b). To each point of this plane there corresponds a cubic polynomial, and to each polynomial there corresponds its three roots, among which there may be equal ones. If the polynomial has coincident roots, then the orbit corresponding to it is irregular. Thus, we obtain the equation of the discriminant of this group in the space $\mathbf{C}^2$ by equating to zero the discriminant of the cubic poly-nomial with coefficients a and b. Thus, the discriminant in the given case is the curve $4a^3 + 27b^2 = 0$, that is, a semicubical parabola. This group generated by reflections is the group corre-sponding to all those semicubical singularities encountered by Huygens.

There are also similar constructions for the discriminants of other groups generated by reflections. Thus, for the group generated by reflections in the planes of symmetry of the icosahedron we obtain a certain surface in three-dimensional space. A theorem that we can prove concerning this is that this surface is diffeomorphic to a surface drawn by l'Hôpital (Fig. 18). While the singularities of the evolvents in a neighbourhood of points of convexity of the curve are singularities of type 3/2, to the points of inflexion there corresponds the singularity of the space of irregular orbits of the icosahedral group.

In order to obtain a surface in three-dimensional space from the family of evolvents of a cubical parabola, we need to move all these evolvents in $\mathbf{R}^3$ into different horizontal planes, namely to lift the evolvent corresponding to the parameter value a to the height a. Figure 18 can therefore be regarded as an image of a surface, and the theorem asserts that this surface is diffeomorphic to the discriminant of the group of symmetries of the icosahedron H_3.

§16. The icosahedron and quasicrystals

The theory of groups generated by reflections is connected with crystallography. Namely, some of these groups preserve the crystal lattice. For example, the group considered above, generated by reflections in three lines in a plane, preserves the hexagonal lattice (Fig. 20). All crystallographic groups have been classified, and everything about them is well known. They correspond to simple Lie algebras (and consequently simple Lie groups). However, the group of the icosahedron does not fall under this classification, since it does not preserve any lattice in three-dimensional

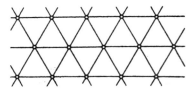

Fig. 20.
A hexagonal crystal lattice

space. In exactly the same way the group of the pentagon does not fall under this classification. This reflects the fact that regular pentagons do not occur in crystals and there are no ornaments with fifth order symmetry than can fill the plane.

Meanwhile in the last few years experimental physicists concerned with the X-ray analysis of crystals have begun to discover pentagons in quite large numbers in X-rays *(23)*. In other words, they have discovered things that apparently have pentagonal symmetry. They have been called quasicrystals, remembering the mathematical theorem that asserts that they cannot be actual crystals.

What can we make of all this? It turns out that in the theory just mentioned – in the theory of singularities connected with evolvents – there is a construction that also leads to similar quasicrystals.

For clarity let us consider not the group of the icosahedron, but the simpler group of symmetries of the pentagon. It is well known that this group cannot be realized as a group of symmetries that preserve some lattice in the plane. Nevertheless, let us consider five-dimensional space, in which, just as in the example considered above, there acts the group of permutations of the coordinates, consisting of 120 elements. This group obviously preserves the five-dimensional lattice of integer points. (In the theory of groups generated by reflections it is known as the group A_4.)

The group of the pentagon, which is embedded in the group of permutations, then also acts in the five-dimensional space, but this representation is reducible. In fact, the rotations of the plane that take a regular pentagon into itself have as eigenvalues the fifth roots of unity. The roots themselves are situated at the vertices of a regular pentagon and, joining the conjugate complex roots in pairs, we find that the space $\mathbf{R}^5$ splits into the sum of three invariant subspaces – a one-dimensional space corresponding to the root 1, and two two-dimensional spaces. The one-dimensional space – the diagonal – is the subspace of vectors whose five coordinates are all equal. In the orthogonal complement to the diagonal there acts, as before, the group of the pentagon, and as before there is a lattice that it preserves. But in the two-dimensional invariant spaces there are no lattices. As simple calculations show, each of the two-dimensional invariant subspaces lies irrationally with respect to the lattice in $\mathbf{R}^4$, that is, it contains no integer point except the origin (this follows from the fact that the golden section is irrational).

Now suppose that in the ambient space there is a function with the symmetry of a lattice (periodic, that is, invariant with respect to shifts by vectors of the lattice and also invariant under the action of all motions and reflections that take the lattice into itself). The restriction of this function to an invariant two-dimensional subspace will not be a periodic function, but an almost-periodic function. This almost-periodic function preserves some rudiments of the symmetry of the pentagon. We can discover them as follows (24).

Let us expand the resulting almost-periodic function in the plane in a Fourier type series, $f(x) = \sum f_k e^{i(k,x)}$. The wave vectors k (the "numbers" of the Fourier harmonics) run through some set of vectors of the dual plane. This set is called the spectrum of the almost-periodic function. In the spectrum there are preserved the traces of the pentagonal symmetry of the original periodic function, defined in the ambient (four-dimensional or five-dimensional) space.

Let us consider first the spectrum of this original periodic function. This spectrum, generally speaking, is an ordinary lattice dual to the original lattice in the ambient space.

To each two-dimensional subspace of any space there corresponds the two-dimensional quotient space of the dual space. It is obtained by factorizing by the space of linear forms equal to zero on the two-dimensional subspace in question.

In other words, the space of wave vectors for the two-dimensional plane in question is obtained from the large space of wave vectors of the ambient space by a natural projection along a subspace of it of codimension two.

In this large space of wave vectors of the ambient space there lies the lattice of Fourier harmonics of the original periodic function,

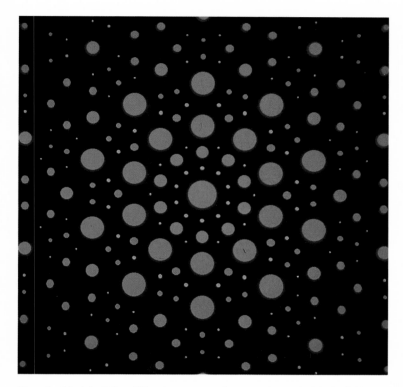

Plate 1. Model of the diffraction pattern for a quasicrystal with fivefold symmetry. The centres of the coloured disks are two-dimensional projections of the integer points in the hyperplane $x_1 + x_2 + x_3 + x_4 + x_5 = 0$ in five-dimensional space. The sizes of the disks decrease exponentially with five-dimensional distance from the origin, and their colours redden with increasing distance from the projection plane (see p. 65).

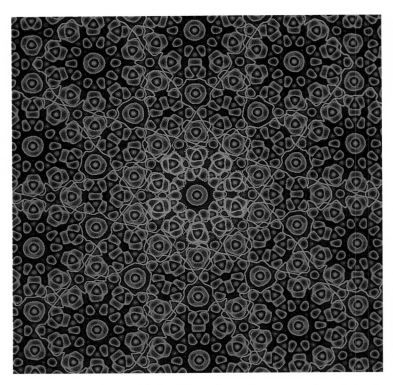

Plate 2a. Stochastic web with five-fold quasicrystallic symmetry. Several chaotic orbits are plotted in blue-green against the gray background of the Hamiltonian flow with the same symmetry (see p. 114 and Plate 3).

Plate 2b. Enlarged section of the map of Plate 2a. The pretzel-shaped chaotic region shown in white is isolated from the main web, shown in blue-green, by a family of invariant curves (see p. 114).

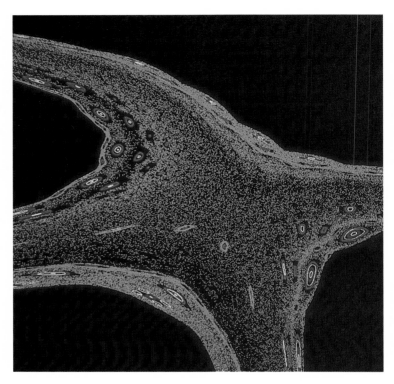

Plate 2c. Enlarged section of the chaotic region of Plate 2b. The fine structure of the chaotic trajectory emerges as well as some islands formed by the quasiperiodic "KAM" trajectories (see p. 114).

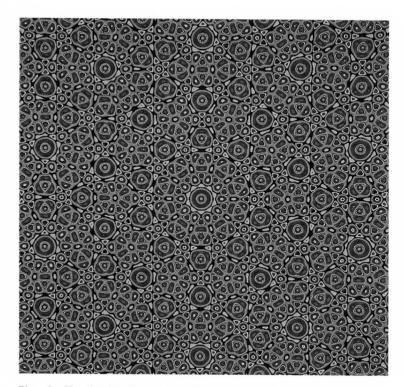

Plate 3. Hamiltonian flow with five-fold quasicrystallic symmetry. Ranges of values of the Hamiltonian H (see pp. 112–114) are assigned different colours, and the boundaries between regions of different colour correspond to phase-space orbits. Seen from a distance, this web looks like a quasiperiodic Penrose tiling.

Plate 4. Quasiperiodic tiling of a portion of the plane by three parallelograms coloured yellow, green and blue. The tiling is obtained by slicing through a three-dimensional cubic lattice with the plane $z=ax+by$, with $a=\sqrt{3}/2$ and $b=\sqrt{2}/2$. The parallelograms are the projections of the cubic faces which intersect the plane (see p. 111).

defined in the ambient space. The spectrum of the almost-periodic function is obtained from this multidimensional lattice by the natural projection of the dual spaces described above (by the projection dual to the embedding of the plane in the ambient space).

In view of the "irrational" position of the two-dimensional plane with respect to the lattice of periods of the ambient space, the projection of the multidimensional space of harmonics is an everywhere dense set in the plane of wave vectors. Thus, the spectrum of the almost-periodic function obtained by restriction to the plane is, generally speaking, an everywhere dense set in the plane of wave vectors. At first glance it is difficult to extract from this any information about the symmetries.

We now turn our attention to the coefficients of the Fourier series. In the original periodic function in the ambient space (which we assume to be smooth) the Fourier coefficients decrease rapidly as the wave vector moves away from the origin. Therefore only finitely many Fourier coefficients corresponding to harmonics with small numbers have an appreciable value.

Consequently, only finitely many harmonics of the almost-periodic function obtained by restriction to the plane have an appreciable value. The corresponding points of the spectrum form a finite set. It is the projection on the plane of a finite cloud of points of the multidimensional lattice close to the origin. This projection preserves the traces of the pentagonal symmetry of the multidimensional lattice in the form of striking (though not quite regular) pentagons (Fig. 21 and Plate 1).

A similar situation exists for the representation of the group of symmetries of the icosahedron (25).

Just this situation explains the connection between the icosahedron and Huygens' evolvents, so the discovery of it can be regarded as the completion of the research begun by Huygens.

On the other hand, in the X-ray analysis of crystals we see essentially the spectrum of a function that has the symmetry of a crystallographic lattice (more precisely, the projection of this three-dimensional spectrum on the plane). Points of the spectrum appear as specks (bright spots) on the X-ray photograph, and the higher the brightness the greater the amplitude of the corresponding harmonics. Therefore in practice we see not the whole spectrum but only the part of it corresponding to harmonics with not very large numbers.

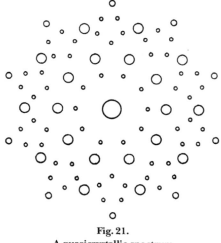

Fig. 21.
A quasicrystallic spectrum

In the analysis of these pictures for certain substances we notice regular structures of approximately regular pentagons (Fig. 21). *(23)*

The theory constructed to explain the connection between icosahedra and evolvents immediately explains how such a spectrum can be obtained. The function in question in three-dimensional space must be not periodic but almost-periodic. Namely, it must be obtained from a periodic function of six variables that admits the symmetry of the icosahedron under restriction to an "irrational" three-dimensional space. Without dwelling on the question of the physical meaning of the three additional "quantum" variables, we just observe that theories proposed by physicists to explain the observations of quasicrystals are close in their ideas to the constructions described above, which arise as a secondary product of the investigation of the singularities of evolvents and Huygens wave fronts – it is one more example of the astonishing unity which so struck Newton and his contemporaries that they interpreted it as proof of the existence of God.

66

CHAPTER 4.
CELESTIAL MECHANICS

§17. Newton after the *Principia*

Neither the discovery of the system of the universe nor the creation of theoretical physics nor the construction of celestial mechanics occurred in the annual plans of Cambridge University or the Royal Society, or in their plans for prospects up to 1700. Newton wrote his 700-page *Principia* in eighteen months at the urgent request of Halley. But since the book was not included in the plan Halley had to publish it on his own account.

At that time Newton was a professor at Trinity College. He had three students. He gave lectures – on arithmetic, geography, optics, and other sciences. His lectures were only given in the Autumn Term (10 lectures a year) and lasted for half an hour. Sometimes there was no audience (Newton's lectures were renowned for their incomprehensibility), and then he just returned home.

Newton spent most of his time and energy on alchemy and theology. His main discoveries were made in his two student years, in the twenty-third and twenty-fourth years of his life. After the *Principia* (which he finished at the age of 44) Newton withdrew from active scientific work.

In 1696 Newton was appointed Warden and then Master of the Mint in London and played a significant part in the economic reforms carried out by his former student Lord Montague Halifax (founder of the Bank of England, and leader of the country during the King's absence).

The revolutionary changes that dragged on for several decades in England, beginning with the Civil War and ending with the "Glorious Revolution" in 1688, brought the economy of the country to a parlous state: corruption and other negative influences of the previous decades required economic reforms, in the course of which it was necessary to withdraw rapidly from circulation the old unsound money, which was not acceptable to foreign states.

In a short time Newton increased the minting of money eightfold, without setting up any new machines. At the same time he set up an investigation and in the one year 1697 he took legal proceedings, as a result of which about 20 forgers were executed.

In 1703 Newton was made President of the Royal Society (see p. 51) and held this office until death in 1727.

§18. The natural philosophy of Newton

Among the most important physical principles contained in the *Principia* we should mention: 1) the idea of relativity of space and time ("in nature there is no body at rest... nor uniform motion"), 2) the conjecture that inertial coordinate systems exist, 3) the principle of determinacy: the positions and velocities of all the particles in the world at an initial instant determine all their future and all their past.

After the *Principia* the universe, which had seemed chaotic, took on the likeness of a well-regulated clock. This regularity and simplicity of the basic principles, from which all the complicated observed motions were derived, were perceived by Newton *(26)* as proof of the existence of God: "This most beautiful system of the sun, the planets, and the comets, could only proceed from the counsel and dominion of an intelligent and powerful being... This being governs all things, not as the soul of the world, but as Lord over all; and on account of his dominion he is wont to be

68

called the Lord God (παντοκρατορ)". Theoretical physics remained in the paradise created by Newton for more than two hundred years, until quantum mechanics and general relativity theory dispelled these illusions.

It is impossible here to list even the main concrete achievements presented in the *Principia*. I recall merely the construction of the theory of limits (which differs from the modern theory only in notation), the topological proof of the transcendence of Abelian integrals (Lemma XXVIII), the calculation of the resistance to motion in a rarefied medium with large supersonic speeds (which found applications only in the age of space travel), the investigation of the variational problem about a body of least resistance for given length and width (the solution of this problem has an internal singularity, of which Newton was aware, but of which his publishers in the 20th century were not apparently aware and smoothed out a figure *(27)*), and the calculation of the perturbations of the motion of the moon by the sun.

§19. The triumphs of celestial mechanics

The development of celestial mechanics after Newton is a long series of triumphs for the law of universal gravitation: the apparent deviations from it with time were accounted for by insufficiently accurate calculation of perturbations. (A notable exception was the precession of the perihelion of Mercury. Its observed value is 599" per century, but calculation of the perturbations gives 557". The discrepancy of 42" is the effect of general relativity theory. New physics usually begins with a refinement of the last digits!)

The first triumph of the theory of gravitation was the prediction of the return of Halley's comet. Halley did not really discover the comet named after him, but noticed the resemblance of the orbits of the comets of 1456, 1531, 1607 and 1682 and ventured to predict the return of the comet after 76 years, that is, in 1758. But because of the perturbation of Jupiter and Saturn the comet was late (according to Clairaut's calculations by 618 days) and crossed the perihelion only in March 1759, almost as Clairaut had predicted.

Another phenomenon that raised doubts about the universality and accuracy of the law of gravitation was the slow but invariably observed acceleration of Jupiter and deceleration of Saturn (Kepler, 1625; Halley, 1695). If this process had been continuing for several million years, it would have completely changed the solar system: Jupiter would have got nearer the Sun, and Saturn further away from it.

The total mass of the planets is roughly a thousandth of the mass of the Sun, so the mutual perturbations of the planets by each other in a year constitute a quantity of the order of a thousandth of the path described. If these perturbations had accumulated over thousands of years, the planets would be able to fall into the Sun or to collide with each other. The Earth would be able to leave the Sun and to be frozen.

Why did this not happen? The reason is that the perturbations experienced by the planets at different times are not actually in the same direction, but have an oscillatory character.

Mathematically such perturbations are expressed as sums of terms proportional to $\cos \omega t$ and $\sin \omega t$ – they are periodic, harmless perturbations. The accumulated perturbations appear in the form of terms increasing with time, proportional to t, or oscillations of increasing amplitude $t\cos(\omega t + \theta)$ (Fig. 22). Terms proportional to time are called *secular*, since, for example, the perturbation at, even if the coefficient a is small (say of the order of one thousandth) becomes large over several centuries.

Thus there arises the problem of secular perturbations: in view of the immensity of cosmological time (billions of years) even very small secular perturbations change the history of the solar system, and in particular the Earth, in a cardinal way.

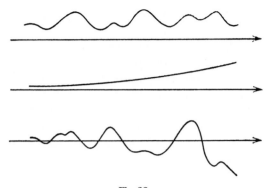

Fig. 22.
Periodic, secular and mixed perturbations

§20. Laplace's theorem on stability

The question is, do there actually exist secular perturbations, or are they artefacts, a consequence of poor mathematical procedure? (For example, let us consider a pendulum oscillating according to the law $x = \cos \omega t$, and suppose we slightly perturb the frequency ω, changing it to $\omega + a$, where a is very small. Then the expansion of the perturbations in a series of powers of a reduces in the first approximation in a to the expression $x = \cos(\omega + a)t = \cos \omega t - at \sin \omega t...$, which contains the dangerous "mixed" term $at \sin \omega t$. Meanwhile the true amplitude of the oscillations of the pendulum does not increase in the course of time, but remains bounded.)

Analysis of planetary perturbations finally led Lagrange (1776) and Laplace (1784) to "Laplace's theorem on the stability of the solar system" *(28)*: the mutual perturbations of the planets, moving in slightly eccentric non-intersecting ellipses almost in the same plane and in the same direction, lead only to almost-periodic oscillations of the eccentricities and of inclinations close to zero, while the distances from the Sun oscillate close to their initial values.

In other words, the major axes of the Kepler ellipses do not have secular perturbations.

Laplace's "theorem" was not proved in the strict sense, since he represented the perturbations by series and proved only the absence of secular terms among the first few terms of the series.

Subsequently the absence of secular and mixed terms was established for all the terms of the series. But the fact that there are no secular terms does not imply that the lengths of the major axes of the Kepler ellipses always remain close to their initial values, since the series themselves diverge (some of their terms are large). The first terms of the series give a good approximation for a restricted time interval, but they do not enable us to judge the behaviour of the orbits in cosmological time.

As for the mutual perturbations of Jupiter and Saturn, as Laplace showed in 1784 they lead to only a long-periodic but not secular change in the eccentricities of the orbits with a period of about 900 years. For the 450 years during which the perturbation accumulates it only manages to move Saturn and Jupiter by less than one degree.

It is very important that the orbits are almost in one plane; if the orbit of the Moon turned through 90°, then the eccentricity of the Moon's orbit under the influence of perturbations from the Sun would begin to increase so rapidly that the Moon would run into the Earth in four years *(29)*.

§21. Will the Moon fall to Earth?

For many years the motion of the Moon remained a very complicated problem, since because of the nearness of the Moon we can easily notice the smallest changes in its motion, and in the expansions one has to take account of terms of a high order of smallness. Already in 1693 Halley noticed that when comparing the observations of eclipses from Arab and ancient sources with modern ones the period of rotation of the Moon, and consequently its orbit, had decreased (the "secular acceleration" was 10" per century).

In 1770 the Paris Academy offered a prize for research on whether the theory of gravitation could explain this phenomenon and whether the decrease in the lunar orbit would lead to the Moon falling to Earth. Euler in a competitive essay considered it "strictly established with unquestionable clarity that the secular inequalities of lunar motion could not be caused by gravitational forces". He explained the acceleration of the Moon by the resistance of the medium, which would finally lead to catastrophe*.

But in 1787 Laplace found the explanation: the long-period oscillations of the eccentricity of the orbit of the Earth under the influence of planetary perturbations. The period of these small oscillations is of the order of several tens of thousands of years, so the effect seems secular.

The oscillations of the eccentricity of the orbit of the Earth are one of the main factors causing the approach of glaciers (because of these oscillations the effective latitude of Leningrad in summer oscillates between the latitudes of Taimir and Kiev in the course of several tens of thousands of years – Milankovich, 1939).

As for the Moon, the explanation of Laplace is only half true: after taking account of the changes in the eccentricity of the Earth's orbit there remains an apparent secular acceleration of the Moon (5" per century) caused, it seems, by tidal friction (according to some estimates, the Bering Sea gives almost half the effect). Under the influence of tidal friction the Moon is all the time receding from the Earth, and the rotation of the Earth is slowing down. The days become twice as long in a time of the

---

* Euler's theory is apparently applicable not to the Moon but to Phobos, a satellite of Mars, which is retarded by its atmosphere and that is why it is accelerating. It would land on Mars in 100 million years, but most likely before that it would be destroyed by tidal forces and transformed into a ring.

order of billions of years (the seasonal oscillations of the extension of the day as a consequence of the redistribution of momentum in the atmosphere and the ocean is a hundred times greater than the annual lengthening of the day because of tidal friction). It is the slowing down of the rotation of the Earth that leads to an apparent acceleration of the Moon *(30)*.

§22. The three body problem

While the problem of the motion of two points was solved by Newton, a precise analytic solution of the problem about the motion of at least three attracting material points under general initial conditions (the three body problem, already posed in the *Principia*) has not only not been found but is in a certain sense impossible *(31)*.

Nevertheless Euler *(32)* had already given some special solutions, for which the mutual position of all three bodies remains constant – all the time the bodies are situated either at the vertices of an equilateral triangle (Lagrange *(33)*) or on a straight line.

These solutions seemed to be a purely mathematical curiosity until (in 1906 and later) "Greeks" and "Trojans" were discovered on the orbit of Jupiter – two groups of small planets which form with the Sun and Jupiter two equilateral triangles (the Trojans move behind Jupiter, and the Greeks outstrip it; Fig. 23).

The solutions of the three body problem corresponding to triangles are stable, at least in the linear approximation, while the solutions for which all three bodies are on one line are automatically unstable and so until recently they have been regarded as practically useless.

However, in the age of space travel the position has changed. A station located at the "point of libration" between the Earth and the Sun corresponding to the Euler solution is in optimal conditions for observing the Sun. This position is unstable, like the

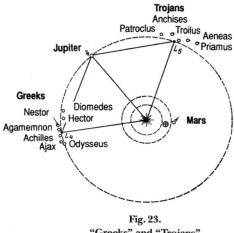

Fig. 23.
"Greeks" and "Trojans"

"head over heels" position of a pendulum. Small random deviations of the station from the point of libration increase in the
course of time. But since the point of libration corresponds to the
exact solution, the rate of growth of small perturbations is small
too. It turns out that the expenditure of energy for a constant
correction of the orbit forcing the station to be close to the point
of libration all the time is small (the less the deviation is regarded
as admissible, the smaller it is). For a proper choice of correction,
taking account of the perturbing influence of other bodies does
not lead to a change in the final deduction *(34)*.

Thus the exact solutions discovered in the 18th century are
now used in practice in space travel.

§23. The Titius-Bode law and the minor planets

At the time of Newton the solar system ended with Saturn.
Uranus was discovered by chance by Herschel on 13 March 1781

("not burdened with traditions, which in the training of experts everywhere restrict the sphere of their duties and the field of admissible activity, he was able to choose unbeaten paths" – Pannekoek, History of Astronomy).

The discovery of Neptune "at the tip of the pen" (in the place predicted by Adams and Leverrier from the perturbations of Uranus) in September 1846 was a new triumph for the law of universal gravitation.

However, the predicted orbit differed greatly from the true one (its mean distance from the Sun was about 30 astronomical units instead of the predicted 38, and the true eccentricity was much less than predicted).

Some investigators think that the discovery of Neptune in the predicted place was a happy accident: existing observations of the perturbations of Uranus imply the predicted position of Neptune only on the basis of the false conjecture adopted by Adams and Leverrier that the radius of the orbit is subject to "Bode's law" (discovered by Titius).

The empirical law of Titius, the publication of which was revealed by Bode in 1772, gives the following dimensions of the orbits: 4 (Mercury), 4 + 3 (Venus), 4 + 3×2 (Earth), 4 + 3×4 (Mars), 4 + 3×8(?), 4 + 3×16 (Jupiter), 4 + 3×32 (Saturn), 4 + 3×64 (Uranus). The number 28 had to be omitted, since there was no known planet between Mars and Jupiter. People began to look for the missing planet.

The discovery on 1 January 1801 of the minor planet Ceres (diameter about 1000 km) was quickly followed by the discovery of Pallas (600 km), Vesta and Juno. The orbits of all these minor planets were found between the orbits of Mars and Jupiter. At the present time astronomers regularly trace the motion of two and a half thousand similar bodies, now called asteroids, whose diameters range from hundreds of kilometers to hundreds of meters. The number of asteroids of size d or more increases, it seems, in inverse proportion to the square of d. It is believed that the total number of asteroids of diameter greater than a kilometer is at least a million.

The orbits of certain asteroids pass close to the orbit of the Earth. Other asteroids, deviating when passing close to Jupiter or Mars, can greatly change their orbits and also appear close to the Earth.

From modern data collisions of the Earth with asteroids of more than half a kilometer in diameter occur at intervals of a hundred thousand years (35).

Craters formed by these collisions have sizes of the order of tens of kilometers (Kaluga stands on one of these craters), and sometimes hundreds of kilometers (close to the mouth of the River Popigai in northern Siberia). A particularly large asteroid could make a hole in the core of the Earth (Whipple suggests that Iceland was formed in this way).

The probability of a collision with the 20-kilometer asteroid Eros (meeting speed 14 km/s) during the next 400 million years is, according to modern estimates, about 1/5, and the diameter of the crater so formed would be about 250 km.

The consequences of a collision with a large asteroid are similar to the consequences of a nuclear war: the atmosphere hardly lowers the speed of the asteroid, and all its kinetic energy is instantaneously emitted at a stroke on the Earth.

Particularly powerful collisions may have ecological significance and may influence the extinction of forms in various continents and even on the whole planet. Thus, the effect of a collision with an asteroid is comparable with the results of man's activities and does not threaten the integrity of the Earth.

§24. Gaps and resonances

The periods of rotation of the majority of asteroids round the Sun are included between the periods of rotation of Mars and Jupiter,

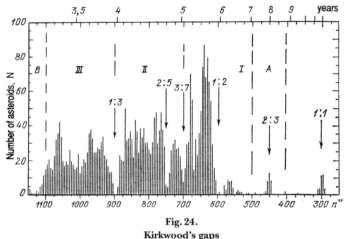

Fig. 24.
Kirkwood's gaps

but fill this interval extremely irregularly. In 1866 Kirkwood discovered "gaps" (Fig. 24) – intervals on the axis of periods free from periods of asteroids. The gaps correspond to resonances (commensurability of the periods): one of the gaps is close to half the period of Jupiter, another is close to 1/3, and there are gaps corresponding to resonances 2/5, 3/5, and so on, the higher the order of the resonance, the smaller is the gap.

There are slits similar to gaps in the ring of Saturn. The largest slit between rings A and B was observed by Cassini in the 17th century. In photographs taken by Voyager-2 *(36)* of ring B of Saturn (Fig. 25) the fine structure is clearly visible: the ring of width 30 000 km consists of a series of thinner rings, separated by wide slits, each of the fine rings is separated by narrower slits into finer rings, and so on, and finally into rings whose width is apparently comparable with its thickness, which is of the order of a kilometer.

The slits in the ring of Saturn correspond to resonances with its satellites. Several years ago in the course of observation from an aeroplane of the occultation of a star by Uranus its rings were ac-

78

Fig. 25.
Ring of Saturn

cidentally discovered. Analysis of their resonance structure
enabled the Soviet astronomers N. N. Gor'kavyi and A. M. Fridman
(37) to predict a whole series of satellites of Uranus. Six months
later, during the flight of Voyager-2 close to Uranus on 24 January
1986, all these satellites were discovered in the predicted positions
from Uranus – one more triumph for Newton's theory of gravitation.

In the hands of Euler, Lagrange and Laplace the mathematical
methods of Newton underwent an enormous technical development, and from the time of Leverrier there has been an excellent
agreement of theory and observations. But as far as ideas were
concerned all these complicated calculations remained versions
of the theory of perturbations created by Newton.

The two hundred year interval from the brilliant discoveries of Huygens and Newton to the geometrization of mathematics by Riemann and Poincaré seems a mathematical desert, filled only by calculations.

Poincaré, the founder of topology and the modern theory of dynamical systems, posed the question anew. Instead of searching for formulae that express the change in the positions of celestial bodies over the course of time, he asked a question about the qualitative behaviour of the orbits: could the planets approach each other, could they fall into the Sun or go far away from it, and so on. Laplace's "theorem" does not give answers to these questions relating to an infinite time interval, since his series, as Poincaré established, diverge.

With his "New methods of celestial mechanics" and "Analysis situs" (topology) Poincaré *(38)* started a new, qualitative, mathematics, about whose applications to celestial mechanics we can say only a few words here.

It turned out that, depending on the initial conditions, motion in a system of three or more bodies is sometimes regular and sometimes chaotic. An example of a regular motion is the planetary motion in an evolving Kepler ellipse, slowly and slightly changing its eccentricity in the course of infinite time and slowly rotating under the action of perturbations, always remaining in a plane that slightly rocks about an unchanging position, as Laplace's approximate theorem predicts.

An example of chaotic motion is the motion of an asteroid close to a Kirkwood gap (J. Wisdom, A.I. Neustadt, J.L. Tennyson, J.R. Cary, D.F. Escande *(39)*) – a resonance interaction with Jupiter leads to "random", chaotic changes in the eccentricity on one side or the other. Successive "jumps" in the eccentricity are weakly dependent. According to the laws of probability theory a chaotically varying eccentricity becomes large, and then the asteroid could, for example, land on Mars. It is suggested that such a mechnism of "casting out" asteroids from the Kirkwood

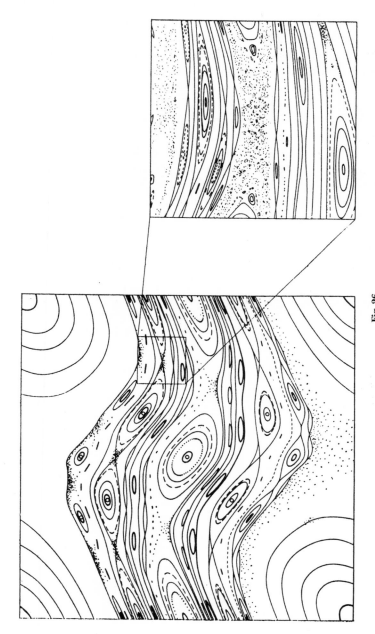

Fig. 26.
Regular and chaotic orbits

81

gaps has led during hundreds of millions of years to the formation of the gaps (Wisdom *(39)*). The orbit of Halley's comet varies chaotically (B.V. Chirikov and V.V. Vyacheslavov *(40)*). The motion of Pluto is chaotic too, according to Wisdom, 1989 (oral report).

The initial conditions of regular and chaotic motions alternate (Fig. 26) like rational and irrational numbers (with the difference that the probabilities of both regular and chaotic behaviour are positive, but the probability that a randomly chosen number is rational is zero). Thus, even if the motion of a planet or an asteroid is regular, an arbitrarily small perturbation of the initial state is sufficient to make it chaotic. Fortunately, however, the rate of development of these chaotic perturbations is extremely small, so the time for which chaos manifests itself under a sufficiently small perturbation of the initial state is large in comparison with the time of existence of the solar system (N.N. Nekhoroshev *(41)*). So for the next billion years the main part of the solar system will hardly change essentially and the "clock mechanism" described by Newton will continue in good working order.

CHAPTER 5.
KEPLER'S SECOND LAW AND THE TOPOLOGY OF ABELIAN INTEGRALS

§25. Newton's theorem on the transcendence of integrals

In the *Principia* there are two purely mathematical pages containing an astonishingly modern topological proof of a remarkable theorem on the transcendence of Abelian integrals *(42)*.

Hidden among research into celestial mechanics, this theorem of Newton has hardly been drawn to the attention of mathematicians. This is possibly because Newton's topological arguments outstripped the level of the science of his time by two hundred years. Newton's proof is essentially based on the investigation of a certain equivalent of the Riemann surfaces of algebraic curves, so it is incomprehensible both from the viewpoint of his contemporaries and also for those twentieth century mathematicians brought up on set theory and the theory of functions of a real variable who are afraid of multivalued functions. Furthermore, Newton was very brief and did not explain many facts that were obvious to him but only entered general mathematical practice later. In addition, having proved a theorem he finally mentioned counterexamples to it that were known to him.

A curve in the plane is said to be algebraic if it satisfies an equation $P(x, y) = 0$, where P is a non-zero polynomial. For example, the circle $x^2 + y^2 = 1$ is an algebraic curve. Other examples of algebraic curves are ellipses, hyperbolas, and the (non-

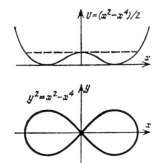

Fig. 27.
The Huygens lemniscate – the algebraic curve $y^2 = x^2 - x^4$ – the energy level curve on the phase plane of a particle moving in the force field of two symmetric potential wells

Bernoulli) lemniscate $y^2 = x^2 - x^4$ (Fig. 27). The sinusoid is not an algebraic curve (why not?).

A function is said to be algebraic if its graph is an algebraic curve. For example, $y = \pm \sqrt{1 - x^2}$ is a two-valued algebraic function.

Let us consider an algebraic oval (a closed convex algebraic curve).

DEFINITION. An oval is said to be *algebraically integrable* if the area of a segment of it can be expressed algebraically.

In other words, the area S of the segment cut off by a line $ax + by = c$ (Fig. 28) must be an algebraic function of the line, that is, it must satisfy an algebraic equation $P(S; a, b, c) = 0$, where P is a non-zero polynomial.

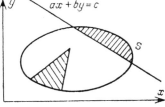

Fig. 28.
The area S as a function of (a, b, c) is non-algebraic

REMARK. If an oval is algebraically integrable, then the area of the sector cut out from it by an angle with vertex lying inside the oval is an algebraic function of the lines forming the angle. This is because the area of the triangle, which is the difference between the sector and the segment, is algebraic (for details see §29 below).

84

Newton set himself the task of finding all algebraically integrable ovals. Here is his result.

THEOREM. *Every algebraically integrable oval has singular points: all smooth ovals are algebraically non-integrable.*

Example. An ellipse is algebraically non-integrable. Hence it follows that Kepler's equation, which determines the position of a planet on the Kepler ellipse as a function of time (in accordance with Kepler's second law, according to which the area swept out by the radius vector is proportional to the time), is transcendental and cannot be solved in algebraic functions.

This example led Newton to his general theorem. It is a surprising theorem, because at first glance there is no obvious connection between algebraic integrability and singular points.

REMARK. In modern notation Kepler's equation has the form $x - e \sin x = t$. This equation plays an important part in the history of mathematics. From the time of Newton the solution x has been sought in the form of a series in powers of the eccentricity e. The series converges when $|e| \leq 0.662743...$

The investigation of the origin of this mysterious constant led Cauchy to the creation of complex analysis.

Such fundamental mathematical concepts and results as Bessel functions, Fourier series, the topological index of a vector field, and the "principle of the argument" of the theory of functions of a complex variable also first appeared in the investigation of Kepler's equation.

Proof of Newton's Theorem. We choose a point 0 inside an oval and rotate a ray issuing from it. If the oval is algebraically integrable, then the area swept out by the radius vector of a point of the oval (Fig. 29) must be an algebraic function of the tangent t of the angle of inclination of the ray to the x-axis.

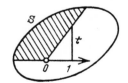

Fig. 29.
The area swept out by the radius vector as a function of t is not algebraic

85

Let us force the ray to run round the oval again and again. On each circuit the area swept out will be increased by the whole value of the area bounded by the oval. Consequently, the area swept out, regarded as a multivalued function of t, has infinitely many different values for the same position of the ray.

But an algebraic function cannot be infinitely multivalued, since the number of roots of a non-zero polynomial cannot exceed its degree.

Consequently, the area swept out is not an algebraic function, and so the oval is not algebraically integrable.

Newton remarked that this argument proves that the length of an arc of an oval is not algebraic.

§26. Local and global algebraicity

Thus, are algebraically integrable ovals non-existent? No, Newton already knew examples of ovals for which the areas of segments are expressed algebraically, and he recalled them in the discussion of his theorem in the *Principia*.

The simplest example is the oval of Fig. 30, $y^2 = x^2 - x^3$. Let t denote the tangent of the angle of inclination of the secant, $y = tx$. Then $t^2 = 1 - x$, so we obtain a parametric representation of the oval,

$$x = 1 - t^2,$$
$$y = t - t^3.$$

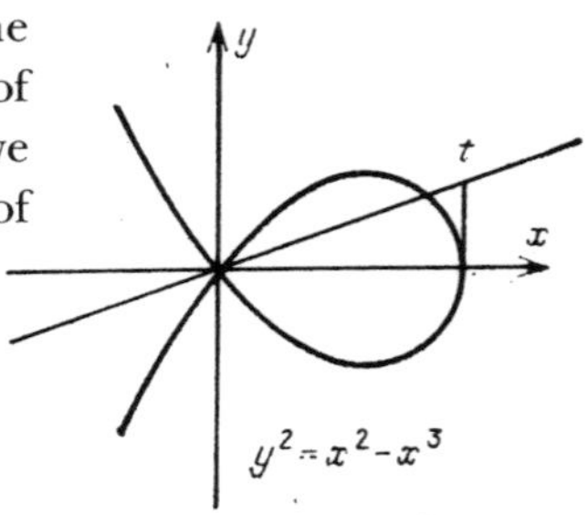

Fig. 30.
A locally algebraically integrable oval with one singular (nodal) point

From this representation it is obvious that the integral for the area $\int y\, dx$ is a polynomial in t. Therefore the area of any segment cut off from this oval by a line can be calculated algebraically.

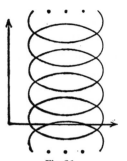

Although the oval thus constructed is not smooth, Newton's arguments can be applied to it. It shows that the area swept out by the radius vector cannot be expressed as a whole by one algebraic function. In fact, each time the ray passes through the singular (nodal) point of the oval the algebraic function that expresses the area swept out jumps to a *new* algebraic function.

Fig. 31.
The graph of a locally algebraic but not algebraic function

The preceding example shows that a function can be *locally* algebraic, even though it is not algebraic as a whole (Fig. 31). In this sense our oval can be called locally algebraically integrable.

In practice local algebraic integrability is almost as useful as the genuine global algebraic integrability. Therefore Newton naturally asked the following question: *can a smooth algebraic oval be locally algebraically integrable?* In other words, can the area S of the segment cut off by the line $ax + by = c$ be an algebraic function of (a, b, c) in a neighbourhood of every point?

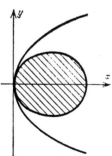

In order to construct (by the method of Fig. 30) a locally algebraically integrable oval that has a tangent everywhere, it is sufficient to choose a suitable pair of polynomials. For example, the polynomials $x = (t^2 - 1)^2$, $y = t^3 - t$ give the oval of Fig. 32, which even has continuous curvature (why?). We have thus obtained an oval which looks like a completely smooth curve, that is, locally algebraically integrable (global algebraic integrability is excluded here by Newton's argument).

Fig. 32.
A locally algebraically integrable oval with continuous curvature

PROBLEM. Construct a (locally) algebraically integrable oval with one singular point which in a neighbourhood of the singular point is the graph of a function having 1989 continuous derivatives (and in the neighbourhoods of the remaining points it coincides with the graphs of infinitely differentiable functions).

§27. Newton's theorem on local non-algebraicity

Thus, a locally algebraically integrable oval can have arbitrarily large finite smoothness (it can be defined everywhere by functions with arbitrarily many derivatives). However, in all our examples there is a singular point on the oval where a derivative of some order is discontinuous.

Newton only regarded as a truly smooth curve one which in the neighbourhood of every point is the graph of a function that can be expanded in a convergent* power series

$$y = a_1 x + a_2 x^2 + a_3 x^3 + \ldots$$

(where the origin is chosen at the point concerned). Such curves are now called *analytic*.

REMARK. The difference in the behaviour of curves of different finite smoothness was well known to Newton and discussed in the *Principia*, and the expansion of all algebraic and elementary functions in rapidly convergent power series was one of his main mathematical achievements.

From the theorem of §25 Newton derived a much more powerful assertion.

THEOREM. *No analytic oval is algebraically integrable, even locally.*

---

* The convergence of the series is not essential for the proof of non-integrability, which works as well in the case where the series are asymptotic.

Proof of local algebraic non-integrability of analytic ovals: If an oval were locally but not globally algebraically integrable, then the area swept out would be expressed by one algebraic function on one side of some point of it and by another algebraic function on the other side. But for an analytic oval the area swept out depends analytically on the direction of the ray. Therefore both these algebraic functions could be expanded in a neighbourhood of this point of the analytic oval in the same convergent power series. Hence both these algebraic functions would coincide in a neighbourhood of the point. But then they would coincide everywhere (this follows from the fact that a polynomial that is not identically zero cannot have more roots than its degree).

Thus, if a locally algebraically integrable analytic oval were to exist, then it would be algebraically integrable globally. But since this is impossible (§25), an analytic oval cannot be algebraically integrable even locally.

§28. Analyticity of smooth algebraic curves

A curve is said to be infinitely smooth if it is locally the graph of a function that is differentiable arbitrarily many times.

THEOREM. *An infinitely smooth algebraic curve is analytic.*

This fact was known to Newton, since he was able to describe the equation of any "branch" of an algebraic curve in a neighbourhood of any point of it in the form of a rapidly convergent series

$$y = a_1 x^{1/n} + a_2 x^{2/n} + a_3 x^{3/n} + \dots$$

(where the origin is placed at the point in question).

[Newton stated the theorem on convergence of this series thus: "The further this result is developed for sufficiently small x, the more it approaches the true value of y, so the difference between it and the exact value of y can eventually be made less than any given quantity" *(43)*.

The series is constructed by means of the Newton polygon; see §8].

Each term of the series with a fractional index has only a bounded number of derivatives. If there is at least one such term with fractional index in the series, then the curve defined by the series cannot be infinitely smooth in a neighbourhood of the point in question.

For an infinitely smooth algebraic curve there are therefore only terms of integral degree in the expansion, and this means that the curve is analytic.

COROLLARY. *No infinitely smooth algebraic oval is algebraically integrable, even locally.*

Thus, an infinitely smooth closed convex curve cannot be even locally algebraically integrable if it is algebraic. Is it possible that non-locally algebraically integrable curves can be found among non-algebraic ovals?

§29. Algebraicity of locally algebraically integrable ovals

A smooth non-algebraic oval is algebraically non-integrable. This follows from what we proved above, since the following theorem is true.

THEOREM. *Any locally algebraically integrable oval is algebraic.*

Newton used this as an obvious fact. Apparently he argued as follows.

LEMMA. *The envelope of any algebraic family of lines is algebraic.*

In other words, if the set of tangents to a curve satisfies an algebraic equation, then the curve itself is algebraic.

Proof of the lemma. Let us consider two neighbouring tangents such that the tangents of their inclinations to the x-axis are t and $t + h$ (Fig. 33).

90

Their point of intersection describes an algebraic curve as t varies
with h fixed (the hatched curve in
Fig. 33). The degree of this curve (that
is, the degree of the polynomial that
defines it) is bounded by a constant
independent of h. (This follows from
the fact that the condition of compati-
bility of two algebraic equations is expressed as the vanishing of a polynomial in their coefficients – a fact discussed by Newton in those two pages of
the *Principia* where he explains at the

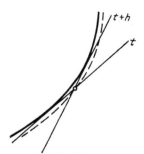

Fig. 33.
The envelope of an alge-
braic family of lines is algebraic

same time that two algebraic curves of degrees m and n intersect
in at most mn points.)

As h tends to zero the point of intersection of neighbouring
tangents tends to the original curve. Thus, since it is the limit of
algebraic curves of bounded degree, the original curve is also
algebraic.

Proof of the theorem. Tangents to an oval cut off from it segments
of zero area. Therefore the tangents $ax + by = c$ to an algebraically
locally integrable oval satisfy an algebraic equation $P(0; a, b, c) =
0$ (see §25). By the lemma the oval is algebraic, which proves the
theorem.

§30. Algebraically non-integrable curves with singularities

Thus, all infinitely smooth ovals are algebraically non-integrable
(even locally). Moreover, Newton's arguments prove the local
algebraic non-integrability of infinitely smooth non-convex
closed non-self-intersecting curves and even many curves with
singularities.

All curves, all of whose singular points are cusps, are locally

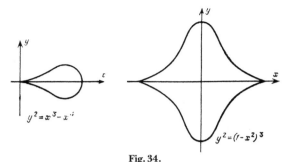

Fig. 34.
Locally algebraically non-integrable curves with cusps

algebraically non-integrable, in particular the curves given by the equations $y^2 = x^3 - x^4$ or $y^2 = (1 - x^2)^3$ (Fig. 34), or curves with singularities of type $y = x^{p/q}$, where q is odd, and so on.

Newton remarked that in order to guarantee local algebraic non-integrability it is sufficient to require that "conjugate branches of a curve going off to infinity" do not approach the points of a closed curve. He obviously had in mind examples like those of Figs. 30 and 32, where there are such "conjugate branches".

In fact, the words "going off to infinity" are put there in error; we need to require that there are no self-intersections. A sufficient condition for a closed curve satisfying an equation $P(x, y) = 0$ to have no self-intersections is that the polynomial P should vanish at exactly two points of a circle with centre at any point of the curve if the radius of the circle is sufficiently small (a more scientific condition for the absence of self-intersections is the following: the oval is in one-to-one correspondence with one of the real connected components of its normalization).

By Newton's method we can prove the following theorem.

THEOREM. *All algebraic curves that are non-self-intersecting in the given sense are algebraically non-integrable (even locally).*

92

On the other hand, a self-intersecting closed curve can certainly be locally algebraically integrable (for some reason Newton overlooked this possibility when he wrote "going off to infinity"). An example is the (non-Bernoulli) lemniscate $y^2 = x^2 - x^4$ (Fig. 27):

$$\int y\, dx = \int x \sqrt{1 - x^2}\, dx = - (\sqrt{1 - x^2})^3 /3$$

is an algebraic function*.

But even for self-intersecting curves algebraic integrability is a rarity.

From Newton's arguments it is clear that the total area bounded by a self-intersecting closed locally algebraically integrable curve (taking account of signs) is zero. For example, the lemniscate is algebraically integrable only because its two loops give opposite contributions to the total area. If we deform the lemniscate so that the absolute values of the areas of the loops become unequal, then it loses the property of being locally algebraically integrable.

§31. Newton's proof and modern mathematics

Newton's theorem may be transferred to hypersurfaces in an even-dimensional space. In an odd-dimensional space things are more complicated. For example, in the three-dimensional case

---

* This example was mentioned by Huygens in a letter to Leibniz in 1691. See also H. Brougham and E.J. Routh, Analytical view of Sir Isaac Newton's *Principia*, London, 1855. Leibniz, in his reply to Huygens, formulated the problem of transcendence of the areas of the segments cut off from an algebraic curve, defined by an equation with rational coefficients, by straight lines with algebraic coefficients (for instance, of the transcendence of the number π and of logarithms of algebraic numbers). The problem of Leibniz is more general than Hilbert's 7th problem, but unlike the last is still, it seems, unsolved.

the value of a spherical segment depends algebraically on the plane cutting it off (Archimedes' theorem). I know of no algebraically integrable bodies other than ellipsoids in odd-dimensional spaces. As V. A. Vasil'ev showed, if there are any, then they must be of very special type. The obvious connections between this question and singularity theory, integral geometry and tomography would probably make it possible to solve it.

Today the ideas on which Newton's proof is based are called the ideas of analytic continuation and monodromy. They lie at the foundation of the theory of Riemann surfaces and a number of branches of modern topology, algebraic geometry and the theory of differential equations, connected above all with the name of Poincaré, those branches where analysis merges with geometry rather than with algebra.

Newton's forgotten *(44)* proof of algebraic non-integrability of ovals was the first "impossibility proof" in the mathematics of the new era – the prototype of future proofs of insolubility of algebraic equations in radicals (Abel) and the insolubility of differential equations in elementary functions or in quadratures (Liouville), and not without reason did Newton compare it with the proof of the irrationality of square roots of integer numbers in the "Elements" of Euclid.

Comparing today the texts of Newton with the comments of his successors, it is striking how Newton's original presentation is more modern, more understandable and richer in ideas than the translation due to commentators of his geometrical ideas into the formal language of the calculus of Leibniz.

APPENDIX 1.
PROOF THAT ORBITS
ARE ELLIPTIC

The proof is based on the next two theorems.

Let us consider an ellipse with centre at the point 0 in the complex plane.

THEOREM 1. *When complex numbers are squared such an ellipse goes into an ellipse with one focus at the point* 0.

For the proof it is convenient to use *Zhukovskii ellipses,* defined by the following construction (Fig. 35).

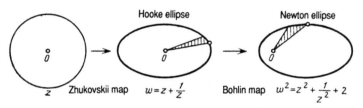

Fig. 35.
The ellipses of Hooke and Newton

LEMMA 1. *When the point z describes a circle $|z| = r > 1$, the point $w = z + 1/z$ describes an ellipse with centre* 0.

Proof of Lemma 1. Let $z = r \cos \varphi + ir \sin \varphi$. Then $w = a \cos \varphi + ib \sin \varphi$, where $a = r + r^{-1}$, $b = r - r^{-1}$, as required: the semiaxes of the Zhukovskii ellipse are equal to a and b.

LEMMA 2. *The foci of a Zhukovskii ellipse lie at the points* ± 2.

Proof of Lemma 2. $c^2 = a^2 - b^2 = 4$.

LEMMA 3. *When complex numbers are squared a Zhukovskii ellipse goes into a Zhukovskii ellipse shifted by* 2.

Proof of Lemma 3. $w^2 = z^2 + 1/z^2 + 2$.

Proof of Theorem 1. The square of a Zhukovskii ellipse has 0 as a focus, since under a shift of 2 the focus -2 goes to 0. Any ellipse with centre at the point 0 is obtained from a suitable Zhukovskii ellipse by a dilatation and a rotation. Hence its square has 0 as a focus.

COROLLARY 1. *Any ellipse with focus 0 is the square of a (unique) ellipse with centre 0.*

Proof of Corollary 1. Among Zhukovskii ellipses there are ellipses with any ratios of semiaxes.

The main feature of the proof of ellipticity of orbits in a gravitational field – the reduction of the motion according to the law of gravitation to a motion according to Hooke's law by squaring the latter motion – is included in the following theorem of Bohlin *(45)*.

THEOREM 2. *Suppose that the point w of the complex plane moves according to Hooke's law $\ddot{w} = -w$. We square w and introduce into the trajectory of the point $Z = w^2$ a new time τ so that the law of areas is satisfied. Then $Z(\tau)$ satisfies the equation of the law of gravitation*

$$\frac{d^2 Z}{d\tau^2} = -\frac{cZ}{|Z|^3}.$$

Proof. From the law of areas $\dfrac{|w|^2 d\varphi}{dt} = \text{const}, \quad \dfrac{2|Z|^2 d\varphi}{d\tau} = \text{const}.$

We choose $\dfrac{d\tau}{dt} = \dfrac{|Z|^2}{|w|^2}.$ Then $\dfrac{d}{d\tau} = \dfrac{w^{-1}\,\overline{w}^{-1}\,d}{dt},$

so

$$\frac{d^2 Z}{d\tau^2} = \frac{1}{w\overline{w}} \frac{d}{dt}\left(\frac{1}{w\overline{w}} \frac{dw^2}{dt}\right) = \frac{2}{w\overline{w}} \frac{d}{dt}\left(\frac{1}{\overline{w}} \frac{dw}{dt}\right) =$$

$$-\frac{2}{w\overline{w}}\left(\frac{1}{\overline{w}^2} \frac{dw}{dt} \frac{d\overline{w}}{dt} + \frac{w}{\overline{w}}\right) = -2w^{-1}\,\overline{w}^{-3}\,(|\dot{w}|^2 + |w|^2) = -4E w^{-1}\,\overline{w}^{-3}$$

($|\dot{w}|^2 + |w|^2 = 2E$ along a trajectory according to the law of conservation of energy). The theorem is proved; $c = 4E$.

96

The ellipticity of motions with negative total energy in the field of a centre attracting according to the usual law of gravitation follows from Theorems 1 and 2 and Corollary 1. In fact, Theorem 2 shows that the squares of Hooke ellipses are orbits of motion in the gravitational field, and Theorem 1 shows that the squares are themselves ellipses with one focus at the attracting centre, and finally from Corollary 1 it is obvious that these squares of suitable Hooke ellipses provide solutions of the equation of the law of gravitation with any preassigned initial conditions, for which the total energy at the initial instant is negative. Since the resulting solutions depend smoothly on the initial conditions, there are no other solutions with these initial conditions.

REMARK. Some mysterious calculations in the proof of Bohlin's theorem possibly become clearer if we slightly generalize the result.

THEOREM 3. *The trajectories of motion of the point w in the complex plane in a central gravitational field whose strength is proportional to the distance from the centre raised to the power a go under the transformation $Z = w^\alpha$ into trajectories of motion in a central field whose strength is proportional to the distance from the centre raised to the power A if*

$$(a + 3)(A + 3) = 4, \qquad \alpha = \frac{a + 3}{2}.$$

Thus, for every power law of attraction there is a unique dual law (Fig. 36). For example, the dual of Hooke's law ($a = 1$) is the law of gravitation ($A = -2$) and conversely. We can derive a formula connecting dual laws from an expression given by Newton for the angle between the pericentres of an almost circular orbit *(46)*. Self-dual laws correspond to $a = -1$ and $a = -5$. These cases were also specially singled out in the *Principia*.

The proof of Theorem 3 repeats the calculation of the proof of Theorem 2:

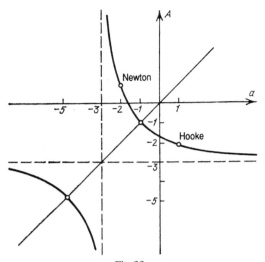

Fig. 36.
Dual laws of attraction

$$\frac{d^2w}{dt^2} = -w\,|w|^{a-1}, \quad \frac{d\tau}{dt} = |w|^{a+1},$$

$$\frac{d^2Z}{d\tau^2} = -CZ\,|Z|^{A-1}, \quad C = 2E\,\alpha\,(\alpha - 1),$$

$$2E = |\dot{w}|^2 + \frac{2\,|w|^{a+1}}{a+1}.$$

COROLLARY 2. *The trajectories of motion in a central field of attraction whose strength is inversely proportional to the fifth power of the distance from the centre are rotated by a suitable inversion.*

Motion in a field whose strength is inversely proportional to the fifth power of the distance had already been considered in the *Principia*: Newton proved that among the trajectories there are circles passing through the centre of attraction *(47)*.

COROLLARY 3. *All trajectories of motion in the usual gravitational field after taking the square root turn into trajectories of motion in the linear central field Z″ = −CZ in the complex plane.*

98

THEOREM 4. *All trajectories of motion in the usual gravitational field are conic sections with one focus at the attracting centre.*

Proof of Theorem 4. Corollary 3 is obtained from Theorem 3 when $a = -2$, $\alpha = 1/2$. Therefore the signs of E and $C = 2E\alpha(\alpha - 1)$ are opposite. The trajectories of motion in the linear field are centrally-symmetric ellipses when $C > 0$, hyperbolas when $C < 0$, and straight lines when $C = 0$. Squaring the ellipses we obtain the Kepler ellipses (by Theorem 1).

When squared a hyperbola with centre at 0 goes into half of a hyperbola with focus at 0. To verify this it is sufficient to consider a Zhukovskii hyperbola: $w = z + 1/z$ describes such a hyperbola when z describes a line passing through 0. Zhukovskii hyperbolas are confocal to Zhukovskii ellipses. When w is squared the focus -2 is shifted to the origin (as for ellipses).

In addition, the second branch disappears, since z^2 describes not a line but a ray.

The case $C = 0$ can be obtained by a limiting process. In addition, it is simpler than those already considered: the line $(t + i)$ when squared goes into the parabola $(t^2 - 1 + 2it)$. Its focus is the point 0, and its directrix is the line $(is - 2)$. This is because

$$(t^2 - 1)^2 + 4t^2 = (t^2 + 1)^2.$$

THEOREM 5. *Under motion in any attracting field whose strength is proportional to the a-th power of the distance from the centre some trajectories are images of straight lines under the transformation*

$$w = Z^\beta, \quad \beta = \frac{1}{\alpha} = \frac{2}{a + 3}.$$

EXAMPLE. In the field of the law of universal gravitation these are parabolic trajectories, $\beta = 2$. In a field whose strength is inversely proportional to the fifth power of the distance from the centre they are circles passing through the centre $\beta = -1$.

Proof. If in the calculation of the proof of Theorem 3 we put $E = 0$, then $d^2 Z / d\tau^2 = 0$, that is, the trajectories Z are straight lines.

REMARK. Let $w(z)$ be any conformal mapping. Then it transforms the trajectories of the motion in the field having potential $U(z) = |dw/dz|^2$ into the trajectories of the motion in the field having the dual potential $V(w) = -|dz/dw|^2$.

The duality of Theorem 3 corresponds to $w = z^\alpha$.

Proof. The duality interchanges E and U in the Maupertuis-Jacobi metrics:

$$\sqrt{2\,(E - U(z))}\ |dz| = \sqrt{E}\sqrt{2(E' - V(w))}\ |dw|, \quad EE' = -1.$$

The duality of Theorem 3 holds also in quantum mechanics (R. Faure, Sur les transformations conformes en mécanique ondulatoire, Comptes rendus Ac. des Sci. Paris, 1953). The quantum duality follows from the evident identity of the quadratic forms

$$\int (\nabla_z \psi)^2 + (a + bU(z))\,\psi^2 d^2z = \int (\nabla_w \psi)^2 + (b - aV(w))\psi^2 d^2w.$$

Of course, similar results hold for any Riemannian kinetic energy metrics, instead of the Euclidean $|dz|$, and even for Finsler metrics.

Examples. The simplest conformal mappings produce the following pairs of dual potentials:

mapping	U	V
$w = 1/z$	$\pm|z|^{-4}$	$\mp|w|^{-4}$
$w = e^z$	$\pm e^{2\,\mathrm{Re}\,z}$	$\mp|w|^{-2}$
$w = \sin z$	$\pm|\cos^2 z|$	$\mp 1/|\sqrt{1 \mp w^2}|$
$w = \tan z$	$\pm|\sec^2 z|$	$\mp 1/|1 \pm w^2|$

Hence the Möbius transformations, which are the only conformal mappings in higher-dimensional spaces, correspond to the Newton case $a = -5$. We see also that the case $a = -3$, whose dual is $A = \infty$, corresponds to "$z^0 \sim \ln z$" and "$r^{-\infty} \sim e^{2x}$", where $\sim$ means "is proportional to" in the formal sense.

APPENDIX 2.
LEMMA XXVIII OF NEWTON'S
PRINCIPIA

There is no oval figure curve whose area, cut off by right lines at pleasure, can be universaly found by means of equations of any number of finite terms and dimensions.

Suppose that within the oval any point is given, about which as a pole a right line is perpetually revolving with an uniform motion, while in that right line a moveable point going out from the pole moves always forward with a velocity proportional to the square of that right line within the oval. By this motion that point will describe a spiral with infinite circumgyrations. Now if a portion of the area of the oval cut out by that right line could be found by a finite equation, the distance of the point from the pole, which is proportional to this area, might be found by the same equation, and therefore all the points of the spiral might be found by a finite equation also; and therefore the intersection of a right line given in position with the spiral might also be found by a finite equation. But every right line infinitely produced cuts a spiral in an infinite number of points; and the equation by which any intersection of two lines is found at the same time exhibits all their intersections by as many roots, and therefore rises to as many dimensions as there are intersections. Because two circles mutually cut one another in two points, one of these intersections is not to be found but by an equation of two dimensions, by which the other intersection may also be found. Because there may be four intersections of two conic sections, any one of them is not to be found universally, but by a an equation of four dimensions, by which they are all found together. For if these intersections are severally sought, because the law and condition of all is the same, the calculus will be the same in every case, and therefore the conclusion always the same, which must therefore comprehend all those intersections at once within itself, and exhibit them all indifferently. Hence it is that the intersections of the conic sections with the curves of the third order, because they may amount to six, come out together by equations of six dimensions; and the intersections of two curves of the third order, because they may amount to nine, come out together by equations of nine dimensions. If this did not necessarily happen, we might reduce all solid to plane problems, and those higher than solid to solid problems *(48)*. But here I speak of curves irreducible in power. For if the equation by which a curve is defined may be reduced to a lower power, the curve will not be one single curve, but composed of two, or more, whose intersections may be severally found by different calculusses. After the same manner the two intersections of right lines with the conic sections come out always by equations of two dimensions; the three intersections of right lines with the irreducible curves of the third order by equations of three dimensions; the four intersections of right lines with the irreducible curves of the fourth order, by equations of four dimensions, and so on *in infinitum.* Wherefore the innumerable intersections of a right line with a spiral, since this is but one simple curve, and not reducible to more curves, require equations infinite in number of dimensions and roots, by which they may all be exhibited together. For the law and calculus of all is the same. For if a perpendicular is let fall from the pole upon that intersecting right line, and that perpendicular together with the intersecting line revolves about the pole, the intersections of the spiral will mutually pass the one into the other; and that which was first or nearest, after one revolution, will be the second; after two, the third; and so on: nor will the equation in the mean time be changed but as the magnitudes of those quantities are changed, by which the position of the intersecting line is determined. Wherefore since those quantities after every revolution return to their first magnitudes, the equation will return to its first form; and consequently one and the same equation will exhibit all the intersections, and will therefore have an infinite number of roots, by which they may all be exhibited. And therefore the intersection of a right line with a spiral cannot be universally found by any finite equation; and of consequence there is no oval figure whose area, cut off by right lines at pleasure, can be universally exhibited by such equation.

By the same argument, if the interval of the pole and point by which the spiral is described is taken proportional to that part of the perimeter of the oval which is cut off, it may be proved that the length of the perimeter cannot be universally exhibited by any finite equation. But here I speak of ovals that are not touched by conjugate figures running out *in infinitum.*

Corollary. Hence the area of an ellipsis, described by a radius drawn from the focus to the moving body, is not to be found from the time given by a finite equation; and therefore cannot be determined by the description of curves geometrically rational. Those curves I call geometrically rational, all the points whereof may be determined by lengths that are definable by equations; that is, by the complicated ratios of lengths. Other curves (such as spirals, quadratrixes, and cycloids) I call geometrically irrational. For the lengths which are or are not as number to number (according to the tenth Book of Elements) are arithmetically rational or irrational. And therefore I cut off an area of an ellipsis proportional to the time in which it is described by a curve geometrically irrational, in the following manner..."

The phrase about the absence of branches going off to infinity was inserted by Newton only in the second edition of 1713. Apparently Newton was unaware of the remarks of Leibniz and Huygens, who had criticised the text of 1687.

"I do not think it possible to ascribe this proposition to Newton, since he never uses any other property of what he calls an oval, but that it is a closed curve that closes after one rotation, which does not even exclude the cases of a square or a triangle", wrote Huygens to Leibniz in 1691 *(49)*.

"Newton, in defending the impossibility of quadrature of an oval, would have replied that such an oval [formed by arcs of two parabolas] is not genuine and does not consist of one curve describing it, as his argument apparently requires, because one of the parabolas does not go into the other when it is extended. But your curve in the form of a figure eight is really describable, and his argument can be applied to it, although it is not at all like an oval, thus, on his argument it cannot be integrable in the general way [to have algebraic areas of segments]. It would be useful to consider his argument in order to understand what is deficient in it. As for a circle or an ellipse, the impossibility of their general integrability has been proved sufficiently, but I do not see any proof of non-integrability of the whole circle or any determined part of it", wrote Leibniz to Huygens on 10/20 April 1691 *(49)*.

A curve in the form of a figure eight, the (non-Bernoulli) lemniscate $y^2 = x^2 - x^4$, was discussed by Huygens in previous letters. Thus, the error in both the original and the revised text of Newton was mentioned by Huygens many years before Newton undertook corrections.

Phrases about the irreducibility of curves were also inserted only in the second edition.

In connection with the proof of Bézout's theorem Newton referred to the fact that otherwise cubic irrationalities would reduce to quadratic, and so on, and it could also be said that he referred to the insolubility of the problem of resolvents or Hilbert's thirteenth problem for algebraic functions (in contrast to Bézout's theorem, these assertions in the general case are still unproved today). In addition, the part of Bézout's theorem necessary to Newton is obvious.

It is incomprehensible why Newton dropped a perpendicular to a moving line: for the proof it is sufficient to restrict oneself to straight lines passing through the central point, and to begin reading from there. Apparently Newton for some reason wished to regard area as a function of a straight line, defined for all lines (despite the opinion that he avoided functions of several variables, here he immediately introduces, in the spirit of Radon transformation, integral geometry or tomography, a function on a manifold of lines).

The connection between the transcendency of functions and the transcendency of numbers, to which Leibniz alluded in the last cited letter to Huygens, is deeper than appears at first sight. In modern times Leibniz's conjecture reads: an Abelian integral along an algebraic curve with rational (algebraic) coefficients taken between limits which are rational (algebraic) numbers is generally a transcendental number. Unlike Hilbert's conjecture on transcendental numbers, which has been proved by Gelfond, this conjecture of Leibniz seems to be still unproved.

NOTES

(1) D. Bennequin, Caustique mystique (d'après Arnol'd et al.), Séminaire Bourbaki 1984/85, Astérisque *133–134* (1986), 19–56.

(2) The malicious Voltaire wrote that Newton owed his career "not to infinitesimal calculus and gravitation but to the beauty of his niece".
Newton's favourite niece Catherine Barton, in whose family he lived for the last twenty years of his life, was celebrated not only for her beauty but also for her intellect. Newton's biographers report that for a long time she was the housekeeper of Newton's pupil Lord Montague Halifax, a poet and distinguished stateman, a member of the regency council of England, First Lord of the Treasury and founder of the Bank of England. After his death in 1715 Catherine Barton inherited a considerable fortune. Newton owed his office of Master of the Mint to Lord Halifax. See T. L. More, Isaac Newton. A biography, Charles Scribner and Sons, New York–London, 1934.

(3) This law is also known as Boyle's law. Boyle actually first published it in 1660 in his book, but with a reference to Hooke as the author of the law, and he did not even pretend to co-authorship. See I. B. Cohen, Newton, Hooke and "Boyle's law" discovered by Power and Towneley, Nature *204* (1964), 618–621.

(4) However, in 1781 Lagrange wrote to d'Alembert about modern mathematics: "...I also think that the mine has become too deep and sooner or later it will be necessary to abandon it if new ore-bearing veins shall not be discovered. Physics and chemistry display now treasures much more brilliant and easily exploitable, thus, apparently, everybody has turned completely in this direction, and possibly posts in geometry in the Academy of Sciences will some day be like chairs in Arabic Language in universities at present".

(5) Later, in 1694, Newton wrote that he had discovered the law of universal gravitation in 1665 or 1666. Still later, in 1714, Newton dated his derivation of the ellipticity of orbits from the inverse square law as 1676 or 1677. However, neither in correspondence with Hooke in 1679 nor earlier did Newton recall his discoveries in this field: he did not publish them and did not speak about them. Newton explained this by the fact that because of the false value of the radius of the Earth he accepted that the calculated accelerations of stones and the Moon do not fit the inverse square law with sufficient accuracy. Hooke's first publication on the force of gravitation as

a possible reason for the ellipticity of orbits was his report read to the Royal Society in 1666, and published in 1674 as part of a 1670 Kutlerian lecture.

(6) It is not difficult to see that Newton's argument gives a deviation of $\omega\sqrt{2h^3/g}$ at a height h above the equator (g is the acceleration due to gravity, and ω is the angular speed of the Earth). A calculation taking account of the Coriolis force, which gives two thirds of this deviation, is given, for example, on p. 131 of the textbook: V. I. Arnol'd, Mathematical methods of classical mechanics, Springer-Verlag, New York, 1989.

(7) Huygens considered systems of colliding and diverging balls joined by strings or rods or rolling in troughs and proved that the centre of gravity of the system never rises above its initial position if we leave the system alone, releasing the balls with no initial speed.

(8) If the graphs of non-coincident analytic functions f and g touch the line $y = x$ at the origin (Fig. 37), then the ratios $|AB|/|BC|$ and $|BC|/|ED|$ tend to one as A tends to the origin. Therefore the required limit of the ratio $|AB|/|D'E'|$ is equal to one.

(9) See. Ch. 6, §7.5 of the book: R. Courant and D. Hilbert, Methods of mathematical physics, Vol. 2, Interscience, New York, 1962. See also: V. I. Arnol'd, On the Newtonian attraction of concentrations of dust particles, Uspekhi Mat. Nauk *37*: 4 (1982), 125; V. I. Arnol'd, On the Newtonian potential of hyperbolic shells, Trudy Tbiliss. Univ. *232–233* (1982), 23–28; A. B. Givental', The polynomial property of electrostatic potentials, Uspekhi Mat. Nauk *39*:5 (1984), 253–254; V. I. Arnol'd, Magnetic analogues of the Newton and Ivory theorems, Uspekhi Mat. Nauk *38*:5 (1983), 145–146; A. D. Vainshtein and B. Z. Shapiro, Multidimensional analogues of the Newton and Ivory theorems, Functional Anal. Appl. *19* (1985), 17–20.

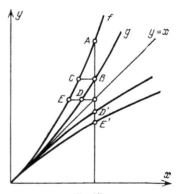

Fig. 37.
Calculation of the limit $|AB|/|D'E'|$

A family of confocal surfaces of the second order in n-dimensional Euclidean space is defined as a family of surfaces dual to the surfaces of a Euclidean pencil of quadrics $(A - \lambda Ex, x) = 1$ with parameter λ. When $n = 2$ the confocal "surfaces" are ellipses and hyperbolas with common foci.

(10) R. Weinstock, Dismantling a centuries-old myth: Newton's *Principia* and inverse-square orbits, Amer. J. Phys. *50* (1982), 610–617.

(11) For example, the equation $\dot{x} = x^{2/3}$ has solutions $x = 0$ and $x = t^3/27$ with common initial condition $x(0) = 0$.

(12) I. Newton, Methodus fluxionum, *Principia*, Liber II, Lemma 2, p. 243.

(13) Before his death Barrow told his friends: "At last I shall know the solutions of many geometrical and astronomical problems. Oh God, what a geometer!"

(14) A. T. Fomenko, A global chronological map, Khimiya i Zhizn' 1983, no. 9, 85–92.

(15) Swift wrote: "Some of my enemies have industriously whispered about that one Isaac Newton, an instrument maker... might possibly pretend to vie with me for fame in future time".

(16) For infinitesimal calculus see N. Bourbaki, Éléments d'histoire des mathématiques, Hermann, Paris, 1960, pp. 178–220.

(17) For modern mathematicians it is generally difficult to read their predecessors, who wrote: "Bob washed his hands" where they should simply have said "There is a $t_1 < 0$ such that the image $\text{Bob}(t_1)$ of the point t_1 under the natural mapping $t \mapsto \text{Bob}(t)$ belongs to the set of people having dirty hands and a t_2 of the half-open interval $(t_1, 0]$ such that the image of the point t_2 under the same mapping belongs to the complement of the set concerned when the point t_1 is considered."

(18) Oeuvres mathématiques de Leibniz, part I, vol. 2, A. Franck, Paris, 1853, p. 255.

(19) Newton was not anti-religious, but rather a secret Arian, a heretic, who denied the dogma of the Trinity. According to his biographers he believed that God could have had other sons, apart from Christ, through whom he revealed his truth to people, and apparently Newton, being also born on 25 December, seriously believed that he was one of these prophets. An interpretation of the Apocalypse and the prophecies of Daniel are due to Newton; in particular, he predicted the fall of the Papal Throne in the year 2000.

(20) A. N. Bogolyubov, Robert Hooke [in Russian], Nauka, Moscow, 1984. On p. 55 of this book Chladni figures, drawn by accumulation of sand near the zeros of an eigenfunction of an oscillating horizontal plate (and discovered by Hooke more than a hundred years before Chladni) are called Lissajou figures (these have apparently still not been discovered in the works of Hooke).

(21) O. V. Lyashko, Classification of critical points of functions on a manifold with a singular boundary, Functional. Anal. Appl. *17* (1983), 187–193; O. P. Shcherbak, Singularities of a family of evolvents in a neighbourhood of a point of inflection of a curve, and the group H_3 generated by reflections, Functional Anal. Appl. *17* (1983), 301–303; O. P. Shcherbak, Wave fronts and reflection groups, Russian Math. Surveys *43*:3 (1988), 149–194; A. B.

Givental', Singular Lagrangian manifolds and their Lagrangian mappings, Itogi Nauki i Tekhniki, Sovremennye Problemy Matematiki, Noveishie Dostizheniya *33* (1988), 55–112 (translated in J. Soviet Math.).

(22) The degrees of the invariants are 2, 6 and 10; the invariant of degree 2 is the square of the distance from the origin, and the invariants of degrees 6

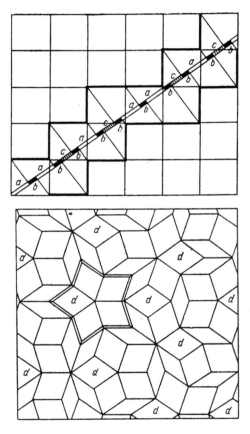

Fig. 38.
The construction of the one-dimensional Penrose tiling and a two-dimensional quasicrystal

and 10 are obtained from the 12 vertices and 20 faces of the icosahedron (as products of linear functions equal to ±1 at the vertices and the centres of the faces).

(23) See, for example: D. Schechtman, D. Gratias and J.W. Cahn, Microscopic evidence for quasi-periodicity in a solid with long-range icosahedral order, C.R. Acad. Sci. Paris *300* (1985), 909–914; M. Senechal and J. Taylor, Quasicrystals: The view from Les Houches, Mathem. Intelligencer *12*:2 (1990), 54–64.

(24) Another way of seeing the pentagonal symmetry is as follows. Let us consider a "staircase" consisting of cubes of the ambient space with vertices at points of the integer lattice that intersect the irrationally disposed subspace in question. The projection of the boundary of the staircase on this subspace determines a partition of it into polyhedra of finitely many types which, however, do not repeat periodically (Plate 4), the so-called Penrose tiling. Fig. 38 shows a similar tiling of the plane with clear traces of pentagonal symmetry.

Incidentally, the construction of the staircase provides tilings of the plane by different rhombi, which cannot be calculated algorithmically (they cannot be constructed by any computer with a finite program).

In fact, a staircase of cubes in three-dimensional space that intersect an irrationally disposed plane is an infinite polyhedron with square faces in three directions that form two bounding polyhedra above and below the surface – lids. We project the upper lid onto the original irrationally disposed plane along a diagonal of the cube. Three adjacent faces of the cube project into three parallelograms affinely equivalent to three rhombi of the same size having angles of 120° at a common vertex (Fig. 39).

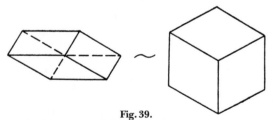

Fig. 39.
On the construction of an uncomputable Penrose tiling

All the faces of the upper lid project into parallelograms equivalent to these three parallelograms under parallel displacement, which simply fill the whole plane. An affine transformation takes all these parallelograms into rhombi.

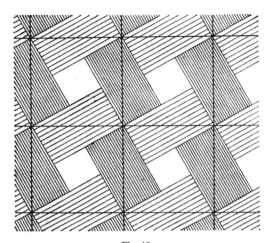

Fig. 40.
**Construction of a quasi-periodic Penrose tiling from a Markov
partition of the torus**

The direction of the original plane is determined by the resulting tiling.
But there is a continuum of directions, while the set of programs is
countable. Hence some of the resulting tilings (and even almost all of
them) are uncomputable.

The Penrose tilings described above are quasi-periodic. Quasi-periodic
Penrose tilings are obtained from a paritition of a torus into prisms with
bases parallel to the irrational subspace on which the quasi-periodic tiling
is cut.

Such a partition of the two-dimensional torus is shown in Fig. 40 (similar
partitions occur in ergodic theory generating the so-called Markov partitions, studied by Adler, Weiss, and Ya. G. Sinai).

The Penrose tilings constructed above (by means of "staircases") are also
cut out from partitions of tori into prisms with parallel bases. So the
construction with prisms can be regarded generating a generalization of
the construction with staircases.

A quasi-periodic Penrose tiling is a partition of an irrational subspace into
level sets of a special quasi-periodic function with finitely many values. This
function is obtained by the restriction to the irrational subspace of a
function with finitely many values on a torus, constant on prisms with bases
parallel to the subspace. We call such quasi-periodic functions Penrose

112

Fig. 41.
A stochastic web

functions. Any continuous function on a torus of any dimensionality can be approximated arbitrarily closely by a function with finitely many values, constant on prisms with bases parallel to a given direction (of any dimensionality). Therefore any quasi-periodic function can be approximated arbitrarily closely by a quasi-periodic Penrose function of the same periodicity (cut out by the same irrational subspace as the original function).

Hence it is clear that the patterns we see when we investigate a quasi-periodic function (its level lines, net of singular points, and so on) must always recall a Penrose tiling. If the original function has any symmetry, then the approximating Penrose tiling (constructed by some natural algorithm) will have the same symmetry.

We thus obtain one more way of generating quasi-crystalline structures: it is sufficient to start with a quasi-periodic function with the necessary symmetry and convert it into a tiling by some natural algorithm.

The following example of this kind was discovered by G. M. Zaslavskii, M. Yu. Zakharov, R. Z. Sagdeev, D. A. Usikov and A. A. Chernikov (A stochastic web and diffusion of particles in a magnetic field, Soviet Physics JETP *64* (1986), 294–303; The generation of ordered structures with an axis of symmetry from Hamiltonian dynamics, JETP Letters *44* (1986), 451–456) in the analysis of a resonance interaction of particles with a wave in a plasma placed in an external magnetic field.

Let us consider a transformation of the plane into itself given by the formula $T = AB$, where A is a rotation through an angle $2\pi p/q$, and $B(x, y)$ $= (x, y + \varepsilon \sin x)$. A computer experiment shows that for a suitable choice of the initial point the images of this point under multiple repetition of the transformation T fill a network of thin lines (for small ε), a "stochastic web" with symmetry of order q, which looks from a distance similar to a Penrose tiling (Fig. 41 and Plates 2a, 2b, 2c).

The explanation is as follows. When $\varepsilon = 0$ the mapping T^q leaves all points of the plane fixed. Therefore for small ε each point under the action of T^q is shifted by a small distance of order ε. On the other hand, the mappings A and B, and hence T, preserve areas. Therefore the mapping T^q up to small quantities of order ε^2 is a transformation in time ε in the phase flow given by some Hamilton function H by the usual formula

$$\frac{dx}{dt} = \frac{\partial H}{\partial y}, \quad \frac{dy}{dt} = -\frac{\partial H}{\partial x}.$$

Calculations show that the Hamilton function $H(x, y)$ has the form

$$H = \cos \alpha_1 + \ldots + \cos \alpha_q,$$

where α_k is a linear function in the plane, equal to the scalar product of the radius vector of the point (x, y) and the radius vector of the k-th vertex of a regular q-gon with centre at the origin. This function H is quasi-periodic and has symmetry of order q. (H is cut out from a sum of cosines defined on a q-dimensional torus under an embedding a of a two-dimensional plane in a q-dimensional torus in the form of the irrational space of an irreducible representation of the cyclic group of order q.)

The Hamilton function H is the first integral of the system of Hamilton equations. Therefore on repeating the mapping T^q a point of the plane will remain on the same level curve of the function H where the initial point lies (at least in a first approximation of the theory of perturbations with respect to ε).

Thus, the observable stochastic web is close to the level curve of the quasi-periodic function H, which has obvious symmetry of order q. This explains the similarity of the web to a quasicrystalline Penrose tiling with symmetry of order q (see Plate 3).

A direct study of level curves of the function H generates the same quasi-crystalline structures in the plane as a repetition of the transformation T that generates the stochastic web.

(25) A. N. Varchenko and S. V. Chmutov, Finite irreducible groups generated by reflections are the monodromy groups of appropriate singularities, Functional Anal. Appl. *18* (1984), 171–183.

(26) *Principia*, Book III, p. 504 in Motte's translation, published by Daniel Adee, New York, 1846.

(27) *Principia*, Book II (On the motion of bodies). An exhortation (pp. 428–430 of: Collected works of Academician A. N. Krylov, vol. VII, I. Newton, The mathematical origin of natural philosophy [in Russian], Akad. Nauk SSSR, Moscow-Leningrad, 1936). See also pp. 65–78 of: V. M. Tikhomirov, Stories of maxima and minima [in Russian], Nauka, Moscow, 1986.

(28) See, for example, Ch. II of Book IV in: P. S. Laplace, Traité de mécanique celeste, I–V, Paris, 1799–1827.

(29) M. L. Lidov, On the approximate analysis of the evolution of orbits of artificial satellites, in: Problems of motion of artificial celestial bodies [in Russian], Akad. Nauk SSSR, Moscow, 1963, pp. 119–134.

(30) See: Tides and resonances in the Solar System [in Russian], Mir, Moscow, 1975 (translations of: G. J. F. Macdonald, Tidal friction, Rev. Geophys. *2* (1964), 467–541; P. Goldreich and S. Soter, Q in the Solar System, Icarus *5* (1966), 375–389; P. Goldreich, History of the lunar orbit, Rev. Geophys. *4* (1966), 411–439; L. V. Morrison, The secular accelerations of the Moon's orbital motion and the Earth's rotation, The Moon *5* (1972), 253–264).

(31) See, for example, V. V. Kozlov, Integrability and non-integrability in Hamiltonian mechanics, Russian Math. Surveys *38*:1 (1983), 1–76.

(32) Nova Comm. Petropol. 11 (1767), 144–151; see also: A. Wintner, The analytical foundations of celestial mechanics, Princeton Univ. Press, 1941.

(33) J.-L. Lagrange, Oeuvres *6* (1772), 272–292.

(34) P. E. El'yasberg and T. A. Timokhova, Control of a spacecraft's motion in the vicinity of a collinear libration centre in the restricted elliptical three body problem, Cosmic Res. *24* (1986), 391–404.

(35) A. N. Simonenko, Asteroids [in Russian], Nauka, Moscow, 1985.

(36) See the photograph at the end of the book: J. Darius, Beyond vision, Oxford University Press, 1984.

(37) N. N. Gor'kavyi and A. M. Fridman, On the resonance nature of the rings of Uranus determined by its undiscovered satellites, Letters in the Astronomical Journal *11* (1985), 717–720.

(38) H. Poincaré, Oeuvres, vols. I–XI, Gauthier-Villars, Paris, 1928–1956; Collected works, vols. I–III, Nauka, Moscow, 1971–1974 (with modern comments in Russian).

(39) A. I. Neishtadt, Change in adiabatic invariant at a separatrix, Soviet J. Plasma Phys. *12* (1986), 568–573.
J. L. Tennyson, J. R. Cary and D. F. Escande, Change of the adiabatic constant due to separatrix crossing, Phys. Rev. Lett. *56* (1986), 2117–2120.
J. Wisdom, A perturbative treatment of motion near the 3/1 commensurability, Icarus *63* (1985), 272–289.

(40) V. V. Vyacheslavov and B. V. Chirikov, The chaotic dynamics of Halley's comet, Preprint 86–184, Inst. of Nuclear Physics, Novosibirsk, 1986.

(41) N. N. Nekhoroshev, An exponential estimate of the time of stability of nearly-integrable Hamiltonian systems, Trudy Sem. Petrovsk., 1979, no. 5, 5–50.

(42) Lemma XXVIII in the *Principia* (pp. 101–105 of this book).

(43) See the article: De analysi per aequationes infinitas (On analysis by equations unlimited in the number of their terms), The mathematical papers of Isaac Newton, vol. II, Cambridge Univ. Press, 1968, pp. 206–273.

(44) An unconvincing criticism of Newton's proof was given by M. Vygodskii in a note on p. 394 of the Russian translation of: H. G. Zeuthen, Geschichte der Mathematik im XVI und XVII Jahrhundert, Teubner, Leipzig, 1903 (GTTI, Moscow, 1933).

H. W. Turnbull (The mathematical discoveries of Newton, Blackie and Son, London–Glasgow, 1945) notes that Newton's argument "shows distinct traces of the ideas which found their full expression in the theory of groups of both Galois and Lie". The meaning of this phrase, which was drawn to my attention by A. P. Yushkevich, is not entirely clear: what are the Lie groups here? In Zeuthen's book he also discusses a work of Gregory, who proved before Newton that the trigonometric functions are transcendental.

(45) K. Bohlin, Bull. Astr. *28* (1911), 144.

(46) *Principia*, Proposition X.

(47) *Principia*, Proposition VII.

(48) Newton's argument is not clear, but he knew "Bézout's theorem", according to which curves of degrees m and n intersect in at most mn points (or in a whole component). He was able to prove this theorem in the following way. If the equations $f(x, y) = 0$, $g(x, y) = 0$ (deg $f = m$, deg $g = n$) are soluble for x for a given y, then the system $fu + gv = 0$ of $m + n$ linear homogeneous equations in the unknown polynomials $u(x)$ and $v(x)$ of degrees $n - 1$ and $m - 1$ has a non-zero solution. Then some polynomial in the coefficients of the system, called the resultant (or as we now call it, the determinant of the system) vanishes. Direct calculations show that the degree of the resultant in y is equal to mn (prove it!). If the resultant is not identically zero, then the number of its roots y does not exceed mn. The resultant was explicitly discussed by Newton.

Thus, the set of points of intersection is projected onto the y-axis, and hence onto any line, into no more than mn points, so it consists of at most mn points.

Also, Newton uses below only the obvious fact that there are finitely many points of intersection of an algebraic curve with a straight line (and the fact that a spiral is irreducible).

(49) Oeuvres mathématiques de Leibniz, part I, vol. 2, A. Franck, Paris, 1853, pp. 90–93.

INDEX